Tribological Performance of Artificial Joints

Tribological Performance of Artificial Joints

Special Issue Editors

Amir Kamali

J. Philippe Kretzer

MDPI • Basel • Beijing • Wuhan • Barcelona • Belgrade

Special Issue Editors
Amir Kamali
Smith & Nephew Orthopaedics Ltd. Aurora House
UK

J. Philippe Kretzer
Universitätsklinikum Heidelberg
Germany

Editorial Office
MDPI
St. Alban-Anlage 66
4052 Basel, Switzerland

This is a reprint of articles from the Special Issue published online in the open access journal *Lubricants* (ISSN 2075-4442) from 2014 to 2015 (available at: https://www.mdpi.com/journal/lubricants/special_issues/Joints)

For citation purposes, cite each article independently as indicated on the article page online and as indicated below:

LastName, A.A.; LastName, B.B.; LastName, C.C. Article Title. *Journal Name* **Year**, *Article Number*, Page Range.

ISBN 978-3-03921-078-7 (Pbk)
ISBN 978-3-03921-079-4 (PDF)

Cover image courtesy of Linos Kretzer.

Contents

About the Special Issue Editors

Amir Kamali is a biomedical engineer working in the Global Clinical Strategy Division of Smith & Nephew in the United Kingdom. Amir has worked at Smith & Nephew for over 14 years in various R&D roles. He was awarded an Honours degree in Medical Engineering from Bradford University/UK in 2001 and a PhD in Bioengineering from the University of Leeds in 2005. Amir has published widely on issues related to biotribology and joint replacement, co-authoring with surgeons and researchers from various organisations. Amongst other responsibilities in the past 5 years, Amir has managed Smith & Nephew's investigator-initiated clinical studies, global clinical publication strategy, S&N's Reference Centers, and the development of a digital platform that offers an innovative approach to patient engagement and patient-reported outcomes data collection.

J. Philippe Kretzer is the Research Director of the Laboratory of Biomechanics and Implant Research of Heidelberg University Hospital, Germany. He graduated in biomedical engineering in 2004, received his PhD in 2008, and was appointed as a professor in 2016 by the University of Heidelberg. He is a senior lecturer in Orthopaedic Biomechanics and specialises in biomechanics and tribology of arthroplasty, and he has written extensively in professional journals and book chapters.

Preface to "Tribological Performance of Artificial Joints"

Joint replacement is a very successful medical treatment. However, the survivorship of the implants could be adversely affected due to the loss of materials in the form of particles or ions as the bearing surfaces articulate against each other. The consequent tissue and immune response to the wear products remains one of the key factors of their failure. Tribology has been defined as the science and technology of interacting surfaces in relative motion and all related wear products (e.g., particles, ions).

Over the last few decades, in an attempt to understand and improve joint replacement technology, the tribological performance of several material combinations has been studied experimentally and assessed clinically. In addition, research has focused on the biological effects and long-term consequences of wear products. Improvements have been made in manufacturing processes, precision engineering capabilities, device designs, and materials properties in order to minimise wear and friction and maximise component longevity in vivo.

This book investigates the in vivo and in vitro performance of orthopaedic implants and their advanced bearings. Contributions are solicited from researchers working in the field of biotribology and bioengineering.

Amir Kamali, J. Philippe Kretzer
Special Issue Editors

Article

Development and Validation of a Wear Model to Predict Polyethylene Wear in a Total Knee Arthroplasty: A Finite Element Analysis

Bernardo Innocenti [1,*], Luc Labey [2], Amir Kamali [3], Walter Pascale [4] and Silvia Pianigiani [4]

1. BEAMS Department (Bio Electro and Mechanical Systems), Université Libre de Bruxelles, Avenue Roosevelt 50, 1050 Bruxelles, Belgium
2. KU Leuven, Mechanical Engineering Technology TC, Kleinhoefstraat 4, 2440 GEEL, Belgium; luc.labey@kuleuven.be
3. Smith & Nephew Orthopaedics Ltd. Aurora House, Harrison Way, Leamington Spa, CV31 3HL, UK; amir.kamali@smith-nephew.com
4. IRCCS Istituto Ortopedico Galeazzi, via R. Galeazzi 4, 20161 Milano, Italy; wal.pascale@gmail.com (W.P.); silvia.pianigiani84@gmail.com (S.P.)

* Author to whom correspondence should be addressed; bernardo.innocenti@ulb.ac.be; Tel.: +32-(0)-2-650-35-31; Fax: +32-(0)-2-650-24-82.

External Editor: Duncan E. T. Shepherd
Received: 21 August 2014; in revised form: 17 October 2014; Accepted: 27 October 2014;
Published: 18 November 2014

Abstract: Ultra-high molecular weight polyethylene (UHMWPE) wear in total knee arthroplasty (TKA) components is one of the main reasons of the failure of implants and the consequent necessity of a revision procedure. Experimental wear tests are commonly used to quantify polyethylene wear in an implant, but these procedures are quite expensive and time consuming. On the other hand, numerical models could be used to predict the results of a wear test in less time with less cost. This requires, however, that such a model is not only available, but also validated. Therefore, the aim of this study is to develop and validate a finite element methodology to be used for predicting polyethylene wear in TKAs. Initially, the wear model was calibrated using the results of an experimental roll-on-plane wear test. Afterwards, the developed wear model was applied to predict patello-femoral wear. Finally, the numerical model was validated by comparing the numerically-predicted wear, with experimental results achieving good agreement.

Keywords: wear; TKA; validated model; FEA; patello-femoral joint

---

1. Introduction

Total knee arthroplasty (TKA) is a surgical procedure to replace the worn-out, native knee joint. In particular, the cartilage-meniscus-cartilage articular surface is replaced by an ultra-high molecular weight polyethylene (UHMWPE) insert in a metal backing for the lower leg component, which moves against a polished CoCrMo component for the upper leg. This combination of materials has been in use since the early 1960s [1]. Although mechanical failure of the UHMWPE insert has been rare in clinical practice, due to its low wear rate [2], studies have continued to show the adverse effects of wear particles in the joint space surrounding implants, which can lead to clinical failure of the implant (which comes loose) or to pain [2–10] and, ultimately, to a revision of the implant. The number of primary implants is exponentially growing, but unfortunately, also the relative number of the revision implants is increasing [3,4]. With the recent trend of rising numbers of total joint replacements being implanted in younger, more active patients, the wear of the UHMWPE bearings has been a large concern, and understanding the wear mechanism has been of utmost importance to ensuring long-term patient satisfaction and implant survival [11,12].

Hence, it is crucial to be able to pre-clinically evaluate the performance of various implant designs and materials and to provide a better understanding of their wear mechanisms. In order to increase the life of total joints, minimizing the wear of UHWMPE has continued to be a goal of material scientists, engineers and clinicians.

The material properties of UHMWPE have been long studied, and the properties that make this polymer suitable as a bearing material arise from its structural and molecular composition. When the UHMWPE bearing surface is in contact with a metal component, such as in TKAs, the surface-to-surface interaction occurs through microscopic interactions between the opposing surface asperities characterized by plastic deformations [13,14].

In the last few decades, there has been an increasing number of tribology studies to understand the problem of wear in TKA components [15–22]. Experimental testing of UHMWPE wear has been conducted in ever wider arrays of machines, loading conditions and on more types of designs over the years, such as pin on disk/plate, roll-on plane and TKA wear simulators [23,24]. Several developments to reduce wear in TKAs were proposed, such as changing the design of the TKAs, the material properties of the polyethylene for the tibial and the patellar inserts (by cross-linking, for example) and with the use of innovative materials, such as oxidized Zr (Smith & Nephew, Memphis, USA) [25] for femoral components.

Wear testing is a crucial step in the design verification process in the industry, yet it is time consuming and expensive, due to low frequency cycles and testing durations of weeks to months [26, 27]. Testing conditions have been prescribed by standards, such as ISO or ASTM, independent of surgical position [28–32], and discrepancies in experimental results exist between testing machines that use force- or displacement-controlled input parameters [33]. To speed the process up, usually pin-on-disk or roll-on-plane [28,34,35] analyses are performed if the research question only concerns the materials that are used for the TKA components. However, if also the design and the position of the TKA components need to be evaluated, dedicated knee wear simulators are used [36–39]. Therefore, each of these devices presents its own advantages and disadvantages; for example, the laboratory evaluation on a simple pin on disk/plate machine is cheap and rapid; however, the results must be viewed with some caution, since the conditions under which the material is tested are drastically simplified. Additionally, knee wear simulators are mainly aimed at analyzing tibio-femoral mechanics and few include also patello-femoral behavior [40,41].

Even with the actual application of these experimental techniques, wear issues still persist. For that reason, an increasing number of *in silico* studies have concentrated their research analyses on TKA contact forces and stresses in line with some *in vitro* tests performed to analyze wear. Computational methods can provide a simplified and efficient solution to predict prostheses behavior in the orthopedics field [42,43].

In an effort to provide efficient implant wear evaluation to augment experimental testing procedures, several computational wear models have been developed based on different techniques based on different wear models. Computer simulation can reduce the time and cost of testing, not only for the orthopedic field. Moreover, once validated, numerical wear models can be also applied in other configurations or loading conditions (mal-alignment or activities other than gait, for example) to investigate the performances of a TKA under less than optimal or severe loading conditions.

In any case, numerical wear simulations of total joint replacement require validation to establish their ability to reproduce wear rates and damage profiles from retrievals or experimental simulators [28]. To the authors' best knowledge, very few published papers report on validated wear models. This number even decreases if we focus our research on the analysis of the patello-femoral joint.

For these reasons, the aim of our work was to develop and to validate a finite element model (FEM) to predict polyethylene wear for TKAs. The wear model is based on Archard's wear model [44], and the study is subdivided into two main work packages: the first is the calibration of the FEM wear model based on experimental roll-on-plane tests; once the FEM wear model is validated, the second step is its use to predict patello-femoral wear during walking cycles, as performed in an experimental wear simulator. Finally, the predicted volumetric patello-femoral wear was compared with the experimental results.

2. Materials and Methods

2.1. Roll-on-Plane: Experimental

Three blocks of UHMWPE (GUR 1020) underwent an experimental roll-on-plane wear test (Figure 1).

The cobalt chromium (CoCr) rolls were sinusoidally loaded with a vertical mean load of 1450 N, a peak-to-peak amplitude of 200 N and a frequency of 1.2 Hz, while they rotated around their symmetry axis with a variable rotation speed (average speed: 0 rad/s; peak amplitude: 0.75 rad; frequency, 1.2 Hz; rotation in phase with the vertical load). Simultaneously, the polyethylene blocks moved back and forth sinusoidally with a peak-to-peak amplitude of 30 mm and a frequency of 1.2 Hz. Their motion was always opposite of the motion of the contact point on the roll. The 6×10^6 cycles were performed while the contact surfaces were immersed in a bovine serum medium simulating human synovial fluid. The wear of each polyethylene block was measured every 500,000 cycles with a profilometer (SURFCOM 1900SD, Zeiss International, Oberkochen, Germany).

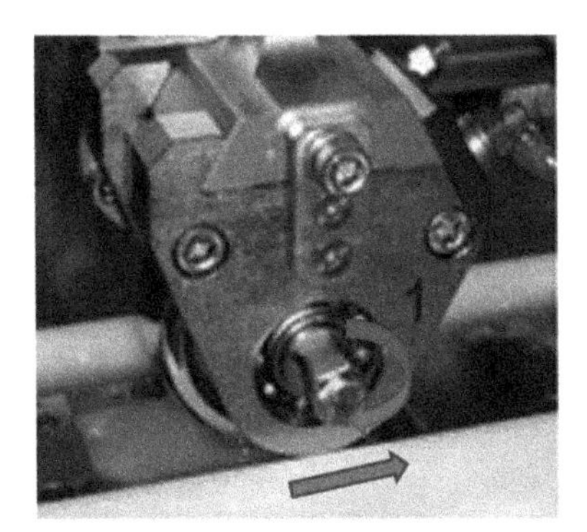

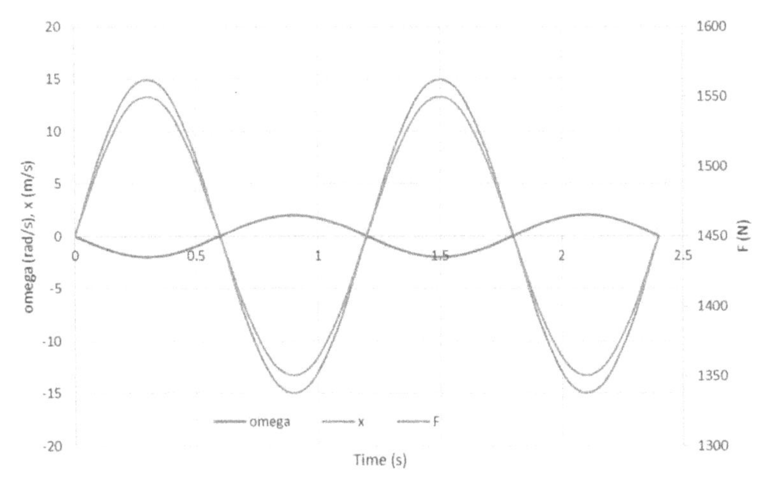

Figure 1. Detail of the roll-on-plane experimental machine.

2.2. *Roll-on-Plane: Numerical Wear Model*

The full experimental roll-on-plane test was reproduced by means of finite element analysis (Figure 2).

For the roller, material properties of CoCr were used with $\rho = 8.27 \times 10^{-3}$ g/mm^3, $E = 240$ GPa and $\nu = 0.3$. For this geometry, 4-mm 10-noded tetrahedral elements were chosen. For the polyethylene block, UHMWPE material properties were used with $\rho = 9.4 \times 10^{-4}$ g/mm^3, $E = 666$ MPa and $\nu = 0.46$. For this geometry, 0.83-mm 8-noded hexahedral elements were used. Both materials are considered linear elastic and isotropic, and a friction coefficient $\mu = 0.05$ was simulated to replicate the experimental conditions. The models were loaded and constrained as in the experimental tests.

Figure 2. Numerical roll-on-plane simulation.

2.3. *Wear Model*

The adhesive/abrasive wear process of UHMWPE was numerically formulated based on the Archard wear model (Archard, 1953) [42]. In 1953, Archard [42] published an equation to estimate the linear wear depth perpendicular to the wear surface of two contacting metal surfaces sliding relative to one another. The equation was known as Archard's wear law and is shown below in Equation 1, in which the linear wear h is determined using the following equation:

$$h = k_W \cdot p \cdot s \tag{1}$$

Where k_W is the wear factor, p is the contact pressure and s is the sliding distance. When contact forces are in the range of those experienced *in vivo*, Archard's law has been shown to reasonably calculate wear depths due to linear sliding of UHMWPE on metal or ceramic [45]. However, the kinematics displayed in total joint replacements are often nonlinear, so the applicability of Archard's law to total joint replacements has been questioned. Moreover, in this expression, delamination, pitting and third body wear are not included, as literature studies report that for UHMWPE, these effects are negligible [46]. To include the friction parameter μ in the model, we adopt the Sakar modification [47] to the Archard model:

$$h = k_W \cdot p \cdot s \cdot (1 + 3\mu^2)^{0.5} \tag{2}$$

The adapted Archard model was used to estimate, after the deformation under a period of cycles, wear and to predict polyethylene geometry modifications due to the wear after a certain number of cycles (CoCr is assumed without modifications). The wear is considered constant for a certain number of cycles (Figure 3).

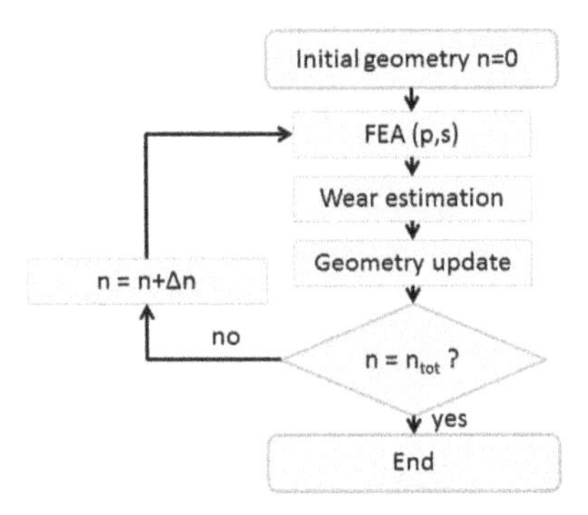

Figure 3. Flow chart of the wear estimation during the FEM modeling.

2.4. Roll-on-Plane Calibration

The wear model is implemented by means of FEM. The simulations were performed with ABAQUS Explicit v6.10 in 3 h 10-min computation time. A Python code was written to implement the wear algorithm in Abaqus, as explained in Figure 3. The geometry of the block was updated every step of 500,000 cycles to reflect the experimental loss of PE material due to wear.

The wear factor was calibrated to fit the numerical prediction to the experimental wear.

2.5. Experimental Patello-Femoral Wear Tests

Three CoCr Genesis II femoral components, Size 5 (Genesis II, Smith & Nephew, Memphis, TN, USA), and the corresponding polyethylene patellar components (32 mm in diameter) underwent experimental tests on a knee wear simulator machine (Figure 4) with simulated 5×10^6 cycles of walking, as reported by Vanbiervliet *et al.* [25]. Patellar flexion, patellar rotation and proximal-distal displacement were derived from the literature on patello-femoral kinematics as a function of the angle of the flexion of the knee. We began the investigation by applying the knee flexion curve from the international standard for wear-testing machines with displacement control (ISO 14243-312); the corresponding patellar flexion, patellar rotation and proximal-distal displacement were then calculated *versus* the cycle time (Figure 5), as reported by Vanbiervliet *et al.* [25].

Figure 4. Experimental patello-femoral test setting.

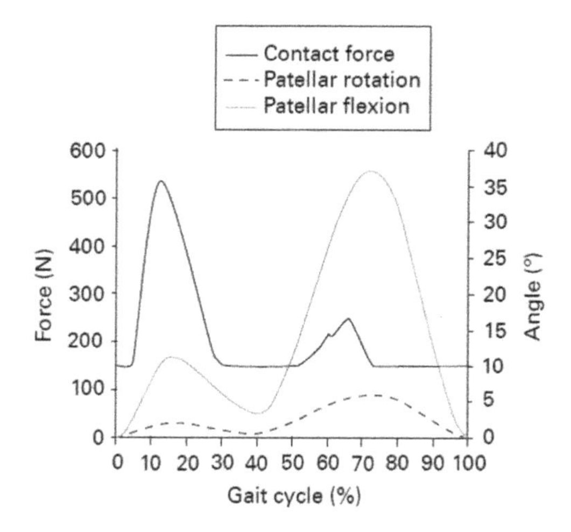

Figure 5. Graphs showing the input curves for the wear simulator.

The volumetric wear was measured, using the weight loss of the components by means of an analytical balance XP205, with an integrated antistatic kit from Mettler-Toledo (Mettler-Toledo International Inc., Greifensee 8606, Zürich, Switzerland).

2.6. Numerical Patello-Femoral Wear Test

Based on the numerical FEM patello-femoral wear analyses performed by Halloran [48] by means of an explicit analysis, the same total knee arthroplasty components have been reproduced in geometries and material properties in FEM (Figure 6) with the same boundary conditions as applied in the wear simulator.

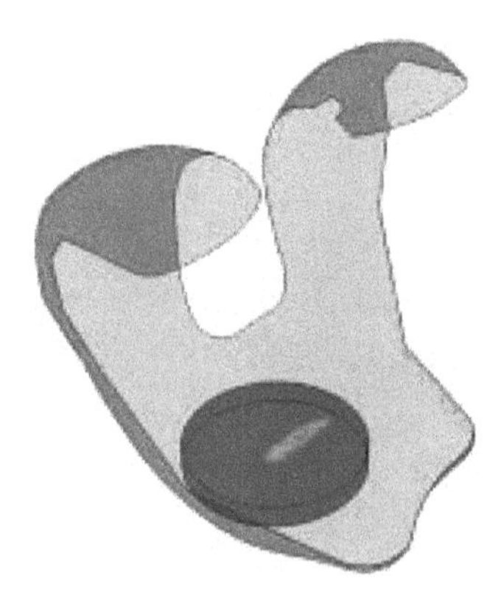

Figure 6. Patello-femoral numerical model.

The material properties of the two components in analysis are the same used for the roll-on-plane simulation, but for this FEM, 1-mm shell elements were adapted for the femoral component and 1-mm hexahedral elements for the patellar component. The applied wear model (Figure 3) is the calibrated one by the roll-on-plane calibration work package.

The predicted volumetric wear volume was compared to the experimental measurements.

3. Results

3.1. Roll-on-Plane

Calibration of the wear model showed that a wear factor of $k_w = 1.83 \times 10^{-8}$ mm^3/Nm gave the best correspondence with the experimental results.

With that wear factor, the wear print for roll-on-plane simulation is in agreement with the experimental one, as shown in Figure 7.

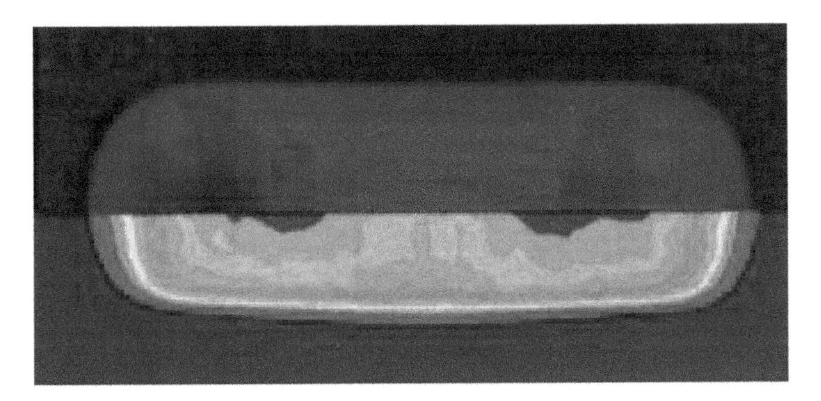

Figure 7. Comparison between experimental and numerical polyethylene wear.

Moreover, with that wear factor value, the FEM results show a maximum linear wear of 0.127 mm, very close to the average maximum depth of the wear track compared to the calculated maximum depth of the wear track (0.125 mm, ±0.01 mm).

3.2. Patello-femoral Test

Results from the experimental test for CoCr femoral components are fully described in Vanbiervliet *et al.* [25].

FEM results show a total volume wear of 0.39 mm^3 after 2×10^6 cycles, in agreement with the mean volume wear measured experimentally for the same number of cycles for three samples, 0.38 ± 0.326.

4. Discussion

The increasing number of tribology studies to analyze polyethylene wear in total knee arthroplasties confirms the need for improved understanding and to find new solutions to avoid the failure of an implant due to polyethylene wear.

Experimental studies are often used, but they are quite expensive and time consuming, and usually, they can analyze only limited configurations and load conditions. For that reason, the use of computational modeling is expanding also in this field, but unfortunately, very few published papers present validated wear models to be used.

Using an FEM analysis, in this study, a model to predict polyethylene wear was developed, calibrated and validated for the wear surface and for the total volumetric wear. The wear model has been applied to predict patello-femoral wear after simulated walking cycles.

The wear model calibration has been performed using data from roll-on-plane experimental tests. Once the model was calibrated, it was applied to predict patello-femoral wear. The results from simulations were compared with some experimental results presenting the same boundary conditions.

FEM computational prediction is in good agreement with the experimental results [25], so the model was validated.

The loading conditions and kinematics were different in the patella-femoral test compared to the roll-on-plane test. While the kinematics in the roll-on-plane test essentially lead to a unidirectional reciprocal motion of the contact between the roll and the plane, this is certainly not true in the patella-femoral experiment [49–52]. Because of the fact that also the rotation of the patellar button is included, there will most certainly occur some cross-shearing in this situation. It has been shown before (although in tibio-femoral testing) that this usually leads to 4–10 times more polyethylene wear [22]. Therefore, in fact, it is rather surprising that the FE model calibrated with the help of the relatively simple roll-on-plane test (without cross-shearing of the polyethylene) does predict also the wear quite well in the more complicated and more realistic patello-femoral wear test, where cross-shearing is present. The reason for this is not entirely clear. One might think that not enough cross-shear was present in our experiment, as Maiti *et al.* have shown that the amount of cross-shear also plays a role in the wear of the replaced patello-femoral joint [41]. In their experiments, wear increased from 8.6 mm^3/MC to 12.3 mm^3/MC, after the amount of patella rotation was increased from 1° to 4°. However, in our experiment, the rotational range of motion was already 5°. Another reason may be that the contact forces and pressures in this set-up are relatively low. The question remains whether the correspondence would still be true in the tibio-femoral articulation, where larger contact forces and pressures are present, but this is the subject of further research.

From the authors' best knowledge, this is the first example of a validated wear model to predict patello-femoral wear. This wear model could also potentially be used for the analysis of wear for tibio-femoral articulations and the analysis of wear for mal-positioned components, such as patellar maltracking. Moreover, this wear model could be implemented in a musculoskeletal system to be able to predict TKA long-term performance for a specific patient. The outputs can be used both for surgeons to better understand the effects of the design and component alignment on wear and for engineers to optimize and improve implant designs.

5. Conclusions

In this study, a numerical procedure to predict polyethylene wear for patello-femoral interactions, after a TKA, by means of finite element analysis was developed and validated. To achieve this, the model was first calibrated on a generic roll-on-plane experimental set up that considers the same material used for TKA. Once the calibration has been performed, the wear model has been used to predict patello-femoral wear under the same boundary conditions of experimental tests. Numerically predicted data have been then compared with experimental outputs founding good agreement. Also comparing with the literature, the developed model assume significance for its use in developing more close-to-real finite elements models that could be used in the orthopaedic clinical and industrial fields in order to help in predicting patients follow-up after a TKA and to improve materials coupled for knee prostheses or TKA designs.

Author Contributions: Bernardo Innocenti designed and performed numerical test, analyzed numerical and experimental data and wrote the paper; Luc Labey and Amir Kamali designed and performed experimental tests, Walter Pascale gave clinical support and conceptual advice and Silvia Pianigiani analyzed numerical and experimental data and wrote the paper. All authors discussed the results and implications and commented on the manuscript at all stages.

Conflicts of Interest: Bernardo Innocenti and Luc Labey were employees of Smith & Nephew. Amir Kamali is an employee of Smith & Nephew. The other authors declare no conflict of interest.

References

1. Charnley, J. Tissue reactions to polytetrafluoroethylene. *Lancet (Int.)* **1963**, *282*, 1379.
2. Wang, A. A unified theory of wear for ultra-high molecular weight polyethylene in multidirectional sliding. *Wear* **2001**, *248*, 38–47. [CrossRef]

3. Johanson, N.A.; Kleinbart, F.A.; Cerynik, D.L.; Brey, J.M.; Ong, K.L.; Kurtz, S.M. Temporal relationship between knee arthroscopy and arthroplasty. A quality measure for joint care? *J. Arthroplast.* **2011**, *26*, 187–191.
4. Curtin, B.; Malkani, A.; Lau, E.; Kurtz, S.; Ong, K. Revision after total knee arthroplasty and unicompartmental knee arthroplasty in the Medicare population. *J. Arthroplast.* **2012**, *27*, 1480–1486. [CrossRef]
5. Blunt, L.; Bills, P.; Jiang, X.; Hardaker, C.; Chakrabarty, G. The role of tribology and metrology in the latest development of bio-materials. *Wear* **2009**, *266*, 424–431. [CrossRef]
6. Matsoukas, G.; Willing, R.; Kim, I.Y. Total hip wear assessment: A comparison between computational and *in vitro* wear assessment techniques using ISO 14242 loading and kinematics. *J. Biomech. Eng.* **2009**, *131*, 1–11.
7. Punt, I.M.; Cleutjens, J.P.M.; de Bruin, T.; Willems, P.C.; Kurtz, S.M.; van Rhijn, L.W.; Schurink, W.H.; van Ooij, A. Periprosthetic tissue reactions observed at revision of total intervertebral disc arthroplasty. *Biomaterials* **2009**, *30*, 2079–2084. [CrossRef] [PubMed]
8. Wroblewski, B.M.; Fleming, P.A.; Siney, P.D. Charnley low-frictional torque arthroplasty of the hip: 20- to 30-year results. *J. Bone Jt. Surg. (Br.)* **2009**, *81*, 427–430. [CrossRef]
9. Sharkey, P.F.; Hozack, W.J.; Rothman, R.H.; Shastri, S.; Jacoby, S.M. Why are total knee arthroplasties failing today? *Clin. Orthop. Relat. Res.* **2002**, *404*, 7–13. [CrossRef]
10. Rand, J.A.; Trousdale, R.T.; Ilstrup, D.M.; Harmsen, W.S. Factors affecting the durability of primary total knee prostheses. *J. Jt. Surg. Am.* **2003**, *85*, 259–265.
11. Kilgour, A.; Elfick, A. Influence of crosslinked polyethylene structure on wear of joint replacements. *Tribol. Int.* **2009**, *42*, 1582–1594. [CrossRef]
12. Carr, B.C.; Goswami, T. Knee implants—Review of models and biomechanics. *Mater. Des.* **2009**, *30*, 398–413. [CrossRef]
13. Wang, A.; Stark, C.; Dumbleton, J.H. Role of cyclic plastic deformation in the wear of UHMWPE acetabular cups. *J. Biomed. Mater. Res.* **1995**, *29*, 619–626. [CrossRef] [PubMed]
14. Wang, A.; Sun, D.C.; Stark, C.; Dumbleton, J.H. Wear mechanisms of UHMWPE in total joint Replacements. *Wear* **1995**, *181–183*, 241–249. [CrossRef]
15. Bartel, D.L.; Bicknell, V.L.; Wright, T.M. The effect of conformity, thickness, and material on stresses in ultra-high molecular weight components for total joint replacement. *J. Bone Jt. Surg. Am.* **1986**, *68*, 1041–1051.
16. Wrona, F.G.; Mayor, M.B.; Collier, J.P.; Jensen, R.E. The correlation between fusion defects and damage in tibial polyethylene bearings. *Clin. Orthop.* **1994**, *299*, 92–103. [PubMed]
17. Blunn, G.W.; Walker, P.S.; Joshi, A.; Hardinge, K. The dominance of cyclic sliding in producing wear in total knee replacements. *Clin. Orthop.* **1991**, *273*, 253–260. [PubMed]
18. Green, T.R.; Fischer, J.; Matthews, L.B.; Stone, M.H. Effect of size and dose on bone resorption activity of macrophages by *in vitro* clinically relevant ultra high molecular weight polyethylene particles. *J. Biomed. Mater. Res.* **2000**, *53*, 490–497. [CrossRef] [PubMed]
19. Schmalzried, T.P.; Jasty, M.; Rosenberg, A.; Harris, W.H. Polyethylene wear debris and tissue reactions in knee as compared to hip replacement prosthesis. *J. Appl. Biomater.* **1994**, *5*, 185–190. [CrossRef] [PubMed]
20. Bohl, J.R.; Bohl, W.R.; Postak, P.D.; Greenwald, A.S. The Coventry Award. The effects of shelf life on clinical outcome for gamma sterilized polyethylene tibial components. *Clin. Orthop.* **1999**, *367*, 28–38. [PubMed]
21. Fisher, J.; Al-Hajjar, M.; Williams, S.; Jennings, L.M.; Ingham, E. *In vitro* Measurement of Wear in Joint Replacements: A Stratified Approach for Enhanced Reliability "SAFER" Pre-Clinical Simulation Testing. *Semin. Arthroplast.* **2012**, *23*, 286–288. [CrossRef]
22. McEwena, H.M.J.; Barnetta, P.I.; Bella, C.J.; Farrarb, R.; Augerc, D.D.; Stoned, M.H.; Fisher, J. The influence of design, materials and kinematics on the *in vitro* wear of total knee replacements. *J. Biomech.* **2005**, *38*, 357–365. [PubMed]
23. Walker, P.S.; Blunn, G.W.; Broome, D.R.; Perry, J.; Watkins, A.; Sathasivam, S.; Dewar, M.E.; Paul, J.P. A knee simulating machine for performance evaluation of total knee replacements. *J. Biomech.* **1997**, *30*, 83–89. [CrossRef] [PubMed]
24. Abdelgaied, A.; Liu, F.; Brockett, C.; Jennings, L.; Fisher, J. Computational wear prediction of artificial knee joints based on a new wear law and formulation. *J. Biomech.* **2011**, *44*, 1108–1116. [CrossRef] [PubMed]
25. Vanbiervliet, J.; Bellemans, J.; Verlinden, C.; Luyckx, J.P.; Labey, L.; Innocenti, B.; Vandenneucker, H. The influence of malrotation and femoral component material on patellofemoral wear during gait. *J. Bone Jt. Surg. Br.* **2011**, *93*, 1348–1354. [CrossRef]

26. Zhao, D.; Sakoda, H.; Sawyer, W.G.; Banks, S.A.; Fregly, B.J. Predicting Knee Replacement Damage in a Simulator Machine Using a Computational Model With a Consistent Wear Factor. *J. Biomech. Eng.* **2008**, *130*, 1–10. [CrossRef]
27. Dunn, A.C.; Steffens, J.G.; Burris, D.L.; Banks, S.A.; Sawyer, W.G. Spatial geometric effects on the friction coefficients of UHMWPE. *Wear* **2008**, *264*, 648–653. [CrossRef]
28. Kang, L.; Galvin, A.L.; Fisher, J.; Jin, Z. Enhanced computational prediction of polyethylene wear in hip joints by incorporating cross-shear and contact pressure in additional to load and sliding distance: Effect of head diameter. *J. Biomech.* **2009**, *42*, 912–918. [CrossRef] [PubMed]
29. Grupp, T.M.; Yue, J.J.; Garcia, R., Jr.; Basson, J.; Schwiesau, J.; Fritz, B.; Blomer, W. Biotribological evaluation of artificial disc arthroplasty devices: Influence of loading and kinematic patterns during *in vitro* wear simulation. *Eur. Spine J.* **2009**, *18*, 98–108. [CrossRef] [PubMed]
30. Giddings, V.L.; Kurtz, S.M.; Edidin, A.A. Total knee replacement polyethylene stresses during loading in a knee simulator. *Trans. ASME* **2001**, *123*, 842–847. [CrossRef]
31. Ghiglieri, W.A.; Laz, P.J.; Petrella, A.J.; Bushelow, M.; Kaddick, C.; Rullkoetter, P.J. Probabilistic cervical disk wear simulation incorporating cross-shear effects. In Proceedings of the 55th Annual Meeting of the Orthopaedic Research Society, Las Vegas, NV, USA, 22–25 February, 2009.
32. Marquez-Barrientos, C.; Banks, S.A.; DesJardins, J.D.; Fregly, B.J. Increased conformity offers diminishing returns for reducing total knee replacement wear. *J. Biomech. Eng.* **2010**. [CrossRef]
33. Schwenke, T.; Orozco, D.; Schneider, E.; Wimmer, M.A. Differences in wear between load and displacement control tested total knee replacements. *Wear* **2009**, *267*, 757–762. [CrossRef]
34. Bei, Y.; Fregly, B.J.; Sawyer, W.G.; Banks, S.A.; Kim, N.H. The Relationship between contact pressure, insert thickness, and mild wear in total knee replacements. *Comput. Model. Eng. Sci.* **2004**, *6*, 145–152.
35. Turell, M.; Wang, A.; Bellare, A. Quantification of the effect of cross-path motion on the wear rate of ultra-high molecular weight polyethylene. *Wear* **2003**, *255*, 1034–1039. [CrossRef]
36. Kang, L.; Galvin, A.L.; Brown, T.D.; Fisher, J.; Jin, Z.M. Wear simulation of UHMWPE hip implants by incorporating the effects of cross-shear and contact pressure. *Proc. Inst. Mech. Eng.* **2008**, *222*, 1049–1064. [CrossRef]
37. Fregly, B.J.; Sawyer, G.W.; Harman, M.K.; Banks, S.A. Computational wear prediciton of a total knee replacement from *in vivo* kinematics. *J. Biomech.* **2005**, *38*, 305–314. [CrossRef] [PubMed]
38. Walker, P.S.; Blunn, G.W.; Perry, J.P.; Bell, C.J.; Sathasivam, S.; Andriacchi, T.P.; Paul, J.P.; Haider, H.; Campbell, P.A. Methodology for long-term wear testing of total knee replacements. *Clin. Orthop. Relat. Res.* **2000**, *372*, 290–301. [CrossRef] [PubMed]
39. Knight, L.A.; Pal, S.; Coleman, J.C.; Bronson, F.; Haider, H.; Levine, D.L.; Taylor, M.; Rullkoetter, P.J. Comparison of long-term numerical and experimental total knee replacement wear during simulated gait loading. *J. Biomech.* **2007**, *47*, 1550–1558. [CrossRef]
40. Ellison, P.; Barton, D.C.; Esler, C.; Shaw, D.L.; Stone, M.H.; Fisher, J. *In vitro* simulation and quantification of wear within the patellofemoral joint replacement. *J. Biomech.* **2008**, *41*, 1407–1416. [CrossRef] [PubMed]
41. Maiti, R.; Fisher, J.; Rowley, L.; Jennings, L.M. The influence of kinematic conditions and design on the wear of patella-femoral replacements. *Proc. Inst. Mech. Eng. H* **2014**, *228*, 175–181. [CrossRef] [PubMed]
42. Askari, E.; Flores, P.; Dabirrahmani, D.; Appleyard, R. Nonlinear vibration and dynamics of ceramic on ceramic artificial hip joints: A spatial multibody modelling. *Nonlinear Dyn.* **2014**, *76*, 1365–1377. [CrossRef]
43. Askari, E.; Flores, P.; Dabirrahmani, D.; Appleyard, R. Study of the friction-induced vibration and contact mechanics of artificial hip joints. *Tribol. Int.* **2014**, *70*, 1–10. [CrossRef]
44. Archard, J.F. Contact rubbing of flat surfaces. *J. Appl. Phys.* **1953**, *8*, 981–988. [CrossRef]
45. Hegadekattea, V.; Huber, N.; Krafta, O. Modeling and simulation of wear in a pin on disc tribometer. *Tribol. Lett.* **2006**, *24*, 51–60. [CrossRef]
46. Pal, S.; Haider, H.; Laz, P.J.; Knight, L.A.; Rullkoetter, P.J. Probabilistic computational modeling of total knee replacement wear. *Wear* **2008**, *264*, 701–707. [CrossRef]
47. Sarkar, A.D. *Friction and Wear*; Academic Press: London, UK, 1980.
48. Halloran, J.P.; Easley, S.K.; Petrella, A.J.; Rullkoetter, P. Comparison of deformable and elastic foundation finite element simulations for predicting knee replacement mechanics. *J. Biomech. Eng.* **2005**, *127*, 813–818. [CrossRef] [PubMed]

49. Saikko, V. *In vitro* wear simulation on the RandomPOD wear testing system as a screening method for bearing materials intended for total knee arthroplasty. *J. Biomech.* **2014**, *47*, 2774–2778. [CrossRef] [PubMed]
50. Srinivas, G.R.; Deb, A.; Kumar, M.N. A study on polyethylene stresses in mobile-bearing and fixed-bearing total knee arthroplasty (TKA) using explicit finite element analysis. *J. Long-Term Eff. Med. Implant.* **2013**, *23*, 275–283. [CrossRef]
51. Schwiesau, J.; Schilling, C.; Kaddick, C.; Utzschneider, S.; Jansson, V.; Fritz, B.; Blömer, W.; Grupp, T.M. Definition and evaluation of testing scenarios for knee wear simulation under conditions of highly demanding daily activities. *Med. Eng. Phys.* **2013**, *35*, 591–600. [CrossRef] [PubMed]
52. Grupp, T.M.; Saleh, K.J.; Mihalko, W.M.; Hintner, M.; Fritz, B.; Schilling, C.; Schwiesau, J.; Kaddick, C. Effect of anterior-posterior and internal-external motion restraint during knee wear simulation on a posterior stabilised knee design. *J. Biomech.* **2013**, *46*, 491–497. [CrossRef] [PubMed]

Article

Wear Performance of Sequentially Cross-Linked Polyethylene Inserts against Ion-Treated CoCr, TiNbN-Coated CoCr and Al_2O_3 Ceramic Femoral Heads for Total Hip Replacement

Christian Fabry [1,2,*], Carmen Zietz [1], Axel Baumann [2] and Rainer Bader [1]

1. Biomechanics and Implant Technology Research Laboratory, Department of Orthopaedics, University Medicine Rostock, 18057 Rostock, Germany; carmen.zietz@med.uni-rostock.de (C.Z.); rainer.bader@med.uni-rostock.de (R.B.)
2. DOT GmbH, 18059 Rostock, Germany; baumann@dot-coating.de

* Author to whom correspondence should be addressed; christian.fabry@med.uni-rostock.de; Tel.: +49-381-40335-389; Fax: +49-381-40335-99.

Academic Editors: Amir Kamali and J. Philippe Kretzer
Received: 14 January 2015; Accepted: 29 January 2015; Published: 16 February 2015

Abstract: The aim of the present study was to evaluate the biotribology of current surface modifications on femoral heads in terms of wettability, polyethylene wear and ion-release behavior. Three 36 mm diameter ion-treated CoCr heads and three 36 mm diameter TiNbN-coated CoCr heads were articulated against sequentially cross-linked polyethylene inserts (X3) in a hip joint simulator, according to ISO 14242. Within the scope of the study, the cobalt ion release in the lubricant, as well as contact angles at the bearing surfaces, were investigated and compared to 36 mm alumina ceramic femoral heads over a period of 5 million cycles. The mean volumetric wear rates were 2.15 ± 0.18 $mm^3 \cdot million\ cycles^{-1}$ in articulation against the ion-treated CoCr heads, 2.66 ± 0.40 $mm^3 \cdot million\ cycles^{-1}$ for the coupling with the TiNbN-coated heads and 2.17 ± 0.40 $mm^3 \cdot million\ cycles^{-1}$ for the ceramic heads. The TiNbN-coated femoral heads showed a better wettability and a lower ion level in comparison to the ion-treated CoCr heads. Consequently, the low volumes of wear debris, which is comparable to ceramics, and the low concentration of metal ions in the lubrication justifies the use of coated femoral heads.

Keywords: hip joint simulator; titanium niobium nitride; coating; contact angle; ion treatment; cross-linked polyethylene; wear

---

1. Introduction

Since the beginning of low-friction arthroplasty in the 1950s, there has been considerable interest in polyethylene wear and its effect on the long-term survival of total hip replacements. With further developments in the field of sterilization [1,2] and composition, such as cross-linking [3–5] or vitamin E stabilization [6–8], the wear resistance of polyethylene has been extended efficiently during the last decades. Thus, hard-on-soft bearings, in which a femoral head made of ceramic or metal articulates against a polyethylene acetabular component, represent the standard solution in total hip arthroplasty so far.

However, all improvements in polyethylene wear resistance are only of value if the tribological performance of the counterface is optimized, with regard to roughness, wettability and abrasion resistance. Actually, there are several femoral head materials available on the market. Femoral heads made of a cobalt-chromium (CoCr) alloy are commonly used in total hip arthroplasty, owing to their beneficial combination of mechanical strength and ductility. In contrast, their clinical success is limited by the loss of their smooth surface over time, resulting in a greater counterface roughness and accelerated polyethylene wear [9–11]. Popular alternatives to CoCr alloys are oxide ceramics, which are classified to be the reference in the field of hard-on-soft bearings. Significantly increased scratch resistance, improved wettability and a biologically inert behavior rank among the decisive advantages of ceramic materials [12].

In order to increase the surface hardness of standard CoCr heads, without affecting the desired ductility of the substrate, different procedures can be applied. One type of method is ion implantation in which preferably nitrogen ions are embedded into the metal surface under high energy [13]. This procedure results in a phase transformation at the surface, and may lead to hardening of up to a depth of approximately 100 nanometers [14].

Another method to increase the abrasion resistance of CoCr femoral heads is to deposit an external ceramic coating on the metal surface in the range of a few microns, without changing the chemical and mechanical properties of the substrate material. Owing to its barrier effect towards the surrounding tissue, this kind of surface modification is deemed to be one of the preferred solutions for patients with sensitivity to metal ions (e.g., cobalt, nickel and chromium) [15].

However, there are still some concerns around coating delamination and the reduced ion-release behavior with ceramic coatings which are based mainly on outdated studies [16,17]. In the past five years, there has been no *in vitro* study which has investigated the performance of current coatings, with regard to polyethylene wear, ion-release behavior and wettability. Therefore, the aim of this experimental study was to evaluate the effect of two different surface modifications of femoral heads made of cobalt-chromium on wear propagation. For this purpose, titanium niobium nitride (TiNbN) coated CoCr femoral heads, as well as ion-treated (LFIT) CoCr heads, were tested in a hip joint wear simulator. In addition, ion levels in serum and contact angles were determined. The results of the analyses were evaluated and compared with controls based on alumina ceramic (Al_2O_3) heads.

2. Material and Methods

2.1. Test Specimens

Sequentially cross-linked polyethylene inserts (Trident X3, Stryker GmbH & Co. KG, Duisburg, Germany) were combined with 36 mm femoral heads. As acetabular components, suitable Trident PSL 56 mm acetabular cups (Stryker GmbH & Co. KG, Duisburg, Germany) were used. The cross-linking process of the sequentially cross-linked polyethylene insert was performed using compression-molded resin sheets out of GUR 1020 by irradiation with 3 MRads and annealing below the melting temperature, repeated three times alternately [18]. The sequentially cross-linked polyethylene material had a density of 0.9392 g/cm^3, which is used for the calculation of the volumetric wear. Before wear testing, the inserts were pre-soaked in the test liquid used for wear test for 55 days at room temperature. For each combination of sequentially cross-linked polyethylene and femoral head material three running samples and a loaded soak control were used to control the liquid absorption of the inserts.

The polyethylene inserts were combined with 36 mm femoral heads made of Co28Cr6Mo (Stryker GmbH & Co. KG, Duisburg, Germany). Three of the running heads were treated with nitrogen ions (LFIT TM, Stryker GmbH & Co. KG, Duisburg, Germany). In addition, three femoral heads were modified using a titanium niobium nitride coating (TiNbN, DOT GmbH, Rostock, Germany) by strongly poisoned cathode surface technology (SPCS), a special type of physical vapor deposition (PVD) arc deposition technology. In this procedure, the number of inhomogeneities (droplets) in the coating structure is drastically reduced during evaporation. The thickness of the TiNbN coating was 4.5 ± 1.5 µm, which is commonly used in clinically practice.

Furthermore, three 36 mm femoral heads made of alumina ceramic (BIOLOX®*forte*, CeramTec AG, Plochingen, Germany) were used for reference. These ceramic heads were tested as part of a previous wear study [19] which used the same loading scenario. Within the present study, contact angle measurements have been made at these ceramic heads. Furthermore, the lubricant generated within the previous wear test [19] was used to analyze the ion level.

2.2. *Hip Simulator Wear Test*

The wear tests were performed according to ISO 14242 using a six-station hip wear simulator (Endolab GmbH, Rosenheim, Germany). The applied axial load and movements, containing flexion/extension, adduction/abduction and rotation, during one gait cycle, are shown in Figure 1. The tests were performed for 5×10^6 cycles at 1 Hz, in temperature-controlled (37 ± 2 °C) chambers. A lubricant bovine serum (Biochrom AG, Berlin, Germany) with a protein concentration of 30 g/L was used. Ethylenediaminetetraacetic acid (5.85 g/L) and sodium azide (1.85 g/L) were added to the lubricant to prevent the precipitation of metallic ions, and calcium phosphate and bacterial contamination. After every 500,000 cycles the lubricant was changed and wear was detected gravimetrically with a high precision balance (Sartorius ME235S, Sartorius AG, Goettingen, Germany). All samples were changed periodically, every 500,000 cycles, throughout the six stations of the hip simulator. The volumetric wear was calculated by dividing the gravimetrical wear (mg) by the density of the sequentially cross-linked polyethylene (0.9392 g/cm^3). In order to calculate the absorption of the lubricant at the inserts, two further polyethylene inserts were just axially loaded and used as a soak control.

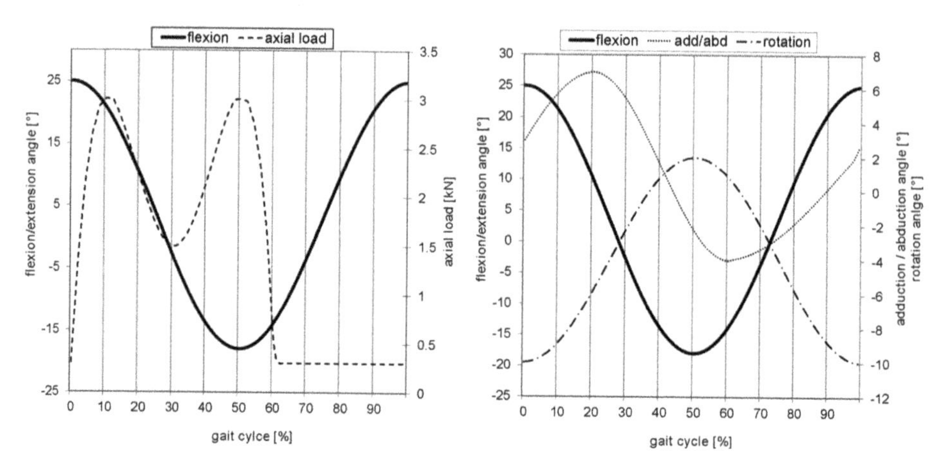

Figure 1. Applied movements throughout the wear simulation as prescribed by the current ISO 14242-1 standard [20].

2.3. *Contact Angle Measurements*

The wetting behavior of the lubricating fluid at the surface of the femoral heads was determined using a drop-shape analyzer (DSA25 Expert, KRÜSS GmbH, Hamburg, Germany). After hip simulator wear testing contact angles were measured at each femoral head in two different areas: first, at the pole of the femoral head representing the main contact area of the bearing; and second, at the inferior area of the femoral head near to the equator, representing a much less stressed articulation area (Figure 2).

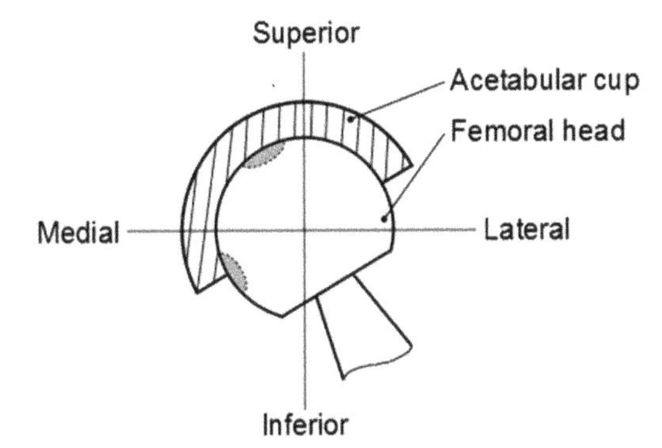

Figure 2. Antero-posterior view of the schematic mounting position according to ISO 14242, with locations of the contact angle measurement area (grey, delimited with dashed lines) at the femoral head.

In the beginning of each measurement the femoral heads were wetted with one droplet of the same lubricant which has been used in the wear simulator test before. Subsequently, an image of the droplet was captured which served as the basis for contact angle analyses. Each measurement at the pole area, as well as at the inferior area of the equator, was applied three times, always using a new droplet. Between the individual droplet analyses, the femoral heads were cleaned in an ultrasonic bath, followed by rinsing in ultrapure-water, and drying at 80 °C for 20 min.

2.4. Analysis of the Cobalt Ion Level

The concentration of cobalt (Co) released from the femoral heads was measured by atomic absorption spectrometry (AAS) (ZEEnit 650, Analytik Jena AG, Jena, Germany) with electrothermal atomization. For detection a cobalt hollow cathode lamp (lamp current: 7.0 mA) emitting light with the wavelength of 240.7 nm was used. Before measurement the AAS was calibrated with well-defined Co concentrations, and the lubricants of the different bearings were diluted to a suitable concentration. The Co-ion level in the lubrication was measured in the lubricant of each bearing, every 500,000 cycles. Subsequently, 20 µL of the diluted lubricant was placed through the sample hole, and onto the platform of the graphite tube from an automated micropipette and sample changer. The tube was heated in a pre-programmed series of steps optimized for Co. The lubricant was evaporated in three steps: 1. 90 °C for 20 s; 2. 105 °C for 20 s; and 3. 110 °C for 10 s. The pyrolize step followed for 10 s at 1000 °C to eliminate residual organic material and to combust the sample into ash. Using a fast heating rate (1300 °C/s) the tube was heated to 2250 °C for about 4 s to vaporize and atomize elements into free atoms. This step included the element analysis. Some of the light emitted by the Co hollow cathode was absorbed in the test chamber by atomized Co atoms. The amount of passed light with the special wavelength was recorded by a detector, and compared with known passed light of adapted concentrations of Co, and thus the ion concentration could be calculated. The tube was cleaned by a final heating step at 2400 °C about 4 s.

2.5. Statistical Analysis

The statistical significance of the volumetric wear of the sequentially cross-linked polyethylene combined with the different femoral heads, the contact angles and Co-ion level of the different bearings, was assessed using the ONEWAY ANOVA test (IBM® SPSS® Statistics version 20 (IBM Corporation, New York, NY, USA)). For comparison of the contact angles at the pole area and at the equator area of the different femoral heads, the independent Student *t*-test was used. The presented data are shown as mean value ± standard deviation. *p*-values of <0.05 were considered significant.

3. Results

3.1. Wear Rates

The wear results for the CoCr and ceramic femoral heads against sequentially cross-linked polyethylene inserts are presented in Figure 3a. All types of femoral heads caused a linear wear behavior of the polyethylene over 5 million cycles without indications of initial bedding-in wear. The polyethylene inserts, combined with nitrogen-treated femoral heads, produced the lowest overall wear with 10 ± 0.88 mm^3, compared to the TiNbN bearings with 13.32 ± 2.00 mm^3.

Based on the overall wear results of this study, the mean wear rates (mm^3 million cycles^{-1}) are demonstrated in Figure 3b. The LFIT CoCr heads produced a polyethylene wear rate of 2.15 ± 0.18 mm^3·million cycles^{-1} in comparison to the TiNbN-coated femoral heads with 2.66 ± 0.40 mm^3 million cycles^{-1}, as well as the alumina ceramic heads with 2.17 ± 0.40 mm^3·million cycles^{-1}. However, the wear rates were not significantly different ($p > 0.05$).

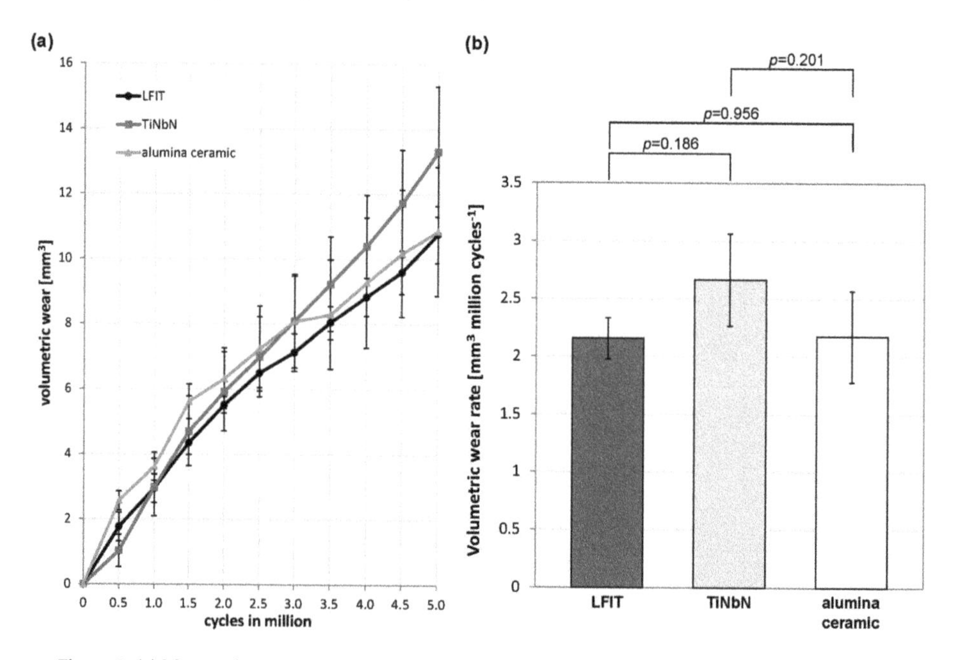

Figure 3. (**a**) Mean volumetric wear and (**b**) wear rates of the sequentially cross-linked polyethylene inserts combined with 36 mm diameter femoral heads modified with nitrogen treatment, TiNbN coating, as well as alumina ceramic [19].

At the end of the hip simulator test, both the CoCr femoral heads as well as the polyethylene inserts showed very small individual scratches on the main contact areas. The TiNbN coatings were undamaged without indications of breakthrough, voids or surface asperities.

3.2. Contact Angle Measurement

The contact angles of the investigated bearing surfaces are shown in Figure 4. The lowest contact angles were determined for the TiNbN coating, followed by angles of the alumina ceramic femoral heads. The differences of the contact angles in the pole area between the different materials were all significant ($p < 0.001$ for LFIT *vs.* TiNbN; LFIT *vs.* alumina ceramic; and TiNbN *vs.* alumina ceramic). At the less stressed equator area, the difference of the angles was not significant for LFIT compared to alumina ceramics ($p = 0.075$). Between LFIT *vs.* TiNbN and TiNbN *vs.* alumina ceramic the angles in the equator area were significantly different: both $p < 0.001$. For the surface-modified femoral heads, the contact angles were significantly higher in the pole area in contrast to the less stressed equator area (LFIT: $p < 0.001$, TiNbN: $p = 0.013$). At the alumina ceramic heads the contact angle was lower in the pole compared with the equator area. This difference was significant ($p = 0.011$).

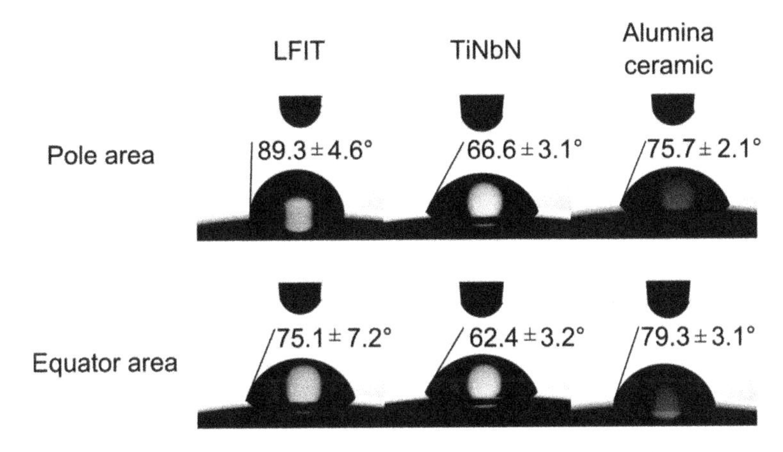

Figure 4. Contact angle measurement at different femoral counterfaces.

3.3. *Cobalt Ion Concentration*

Cobalt ions released into serum during wear testing were detected using atomic absorption spectrometry. The cumulative cobalt concentration after five million cycles was 1511.6 ± 128.2 μg/L for the LFIT femoral heads, 214.5 ± 150.0 μg/L for the TiNbN coupling, and 46.4 ± 4.7 μg/L for alumina ceramic heads. The lubricant of the alumina ceramic bearing demonstrated a small level of cobalt ions, indicating contamination originating from the metallic mountings of the test stations. The overall cumulative Co-ion concentration of the LFIT group was significantly higher than the alumina ceramic ($p < 0.001$) and TiNbN ($p = 0.001$). The difference between alumina ceramic and TiNbN was not significant ($p = 0.191$). Generally, the cobalt ion concentration showed a much larger steady increase for the nitrogen-treated femoral heads compared with the TiNbN specimens (Figure 5). Furthermore, the amount of cobalt ions decreased with the increasing number of cycles for the TiNbN-coated heads.

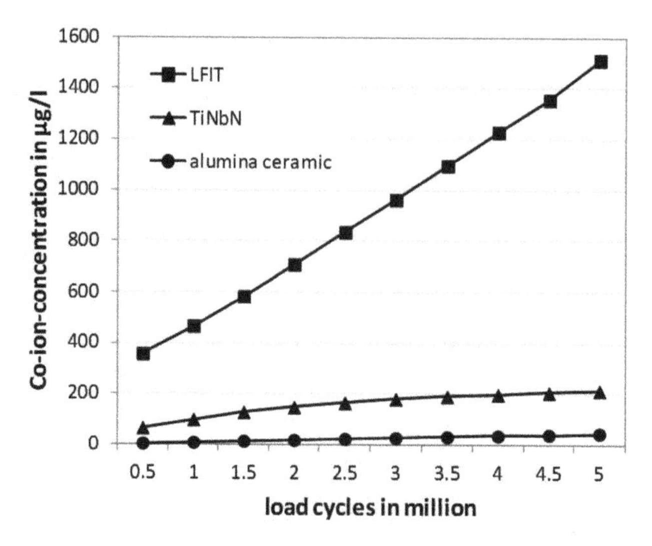

Figure 5. Cumulative cobalt ion concentration over a period of five million cycles analyzed from the used lubricant.

4. Discussion

The aim of this experimental study was to evaluate the influence of two different surface modifications on the wear behavior of sequentially cross-linked polyethylene inserts as well as their effect on the metal ion release.

In total hip arthroplasty, polyethylene wear is one of the major factors limiting the lifetime of the implant inside the human body [21]. Polyethylene wear debris may lead to adverse tissue reactions, followed by extensive bone loss and loosening of the fixation [22,23]. One approach to decrease polyethylene wear is to use CoCr femoral heads with a modified surface.

In our hip simulator wear study, with 36-mm diameter modified CoCr femoral heads against sequentially cross-linked polyethylene inserts, the mean wear rate was 2.15 ± 0.18 mm^3·million cycles^{-1} in combination with the LFIT, and 2.66 ± 0.40 mm^3·million cycles^{-1} for the TiNbN-coated femoral heads. In comparison to previous *in vitro* studies, the results showed that wear rates of both surface modifications were at least three-fold lower than these of traditional 36 mm CoCr-on-cross-linked polyethylene bearings [24–27]. Furthermore, the polyethylene wear could be reduced to the level of alumina ceramic heads [19,26].

In the present contribution, the TiNbN-coated femoral heads demonstrated smooth and intact articulation surfaces without localized damage, such as breakthrough, delamination or cohesive failure over the entire testing period of five million cycles. This excellent wear resistance was consistent with the findings of Galvin *et al.* [28] and Gutmanas *et al.* [29] after hip simulator wear testing with titanium nitride and chromium nitride coated femoral heads against ultra-high-molecular-weight polyethylene. In contrast to the promising *in vitro* results for coatings, some clinical reports showed failures of ceramic coatings in combination with hard-on-soft bearings some years ago. In a case report, Harman *et al.* [16] examined the articular surface of a titanium nitride (TiN) coated CoCr femoral head retrieved after one year of *in-situ* function. Circular voids without TiN coating and surface asperities were evident on the coated femoral heads. In another retrieval study of Raimondi *et al.* [17] fretting and coating breakthrough were observed at 2 out of 4 TiN-coated femoral heads from four patients, 18 to 96 months post-operatively. Both studies concluded an unsafe use of ceramic coatings in the field of hip arthroplasty. The occurred signs of fatigue can be attributed to limitations in the former coating technology, which may have resulted in inhomogeneous layer structures and poor adhesion of the coating. In the past, sputtering was a widely spread process in order to provide a coating at the bearing surfaces, using physical vapor deposition. The purpose was to generate denser coatings with a reduced roughness [13]. However, during sputtering the degree of ionization of the evaporated material is pretty low in comparison to arc deposition (close to 100% right next to the target surface) [30]. The higher the number of ions in the vacuum chamber, the more particles can be accelerated by the bias voltage. Therefore, with arc deposition the particles have a much higher kinetic energy and create coatings with a clearly higher adhesive strength and hardness compared to other deposition methods. In the present study, the TiNbN coating at the femoral heads was applied by a strongly poisoned cathode surface technology (SPCS). In this special type of arc deposition technology, the escape of the reactive gas during physical vapor deposition was guided specifically in order to reduce the number of inhomogeneities (droplets) in the coating structure. Moreover, the surface quality and density achieved with this procedure is comparable to those of the so-called "filtered arc technology", but without its drawbacks such as time- and cost-intensive filter cleaning or low deposition rates.

In addition to the TiNbN-coating, nitrogen treated CoCr femoral heads were used for testing. Similar to the coated femoral heads, a very small number of individual scratches were seen on the main bearing area with the naked eye, indicating that adverse third-bodies influenced the wear testing procedure. Ion treatment with nitrogen and therefore the phase transformation in the microstructure, had a positive impact on the wear behavior of sequentially annealed polyethylene. The surface modification resulted in the lowest average wear rate compared with the TiNbN-coated femoral heads. Nevertheless, differences in wear for the ion-treatment, as well as for the TiNbN-coating, were not statistically significant. In a clinical study, McGrory *et al.* [14] examined roughness and hardness characteristics of retrieved CoCr femoral heads from the same manufacturer, both with and without nitrogen ion treatment. The roughness parameters with ion treatment were lower compared with the non-treated surfaces, indicating that ion treatment increased the scratch resistance of the femoral heads. However, the achieved increased hardness with ion treatment appeared to degrade over time *in vivo* [14].

The secondary purpose of our study was to analyze the wetting behavior of both surface modifications and to compare them with values from alumina ceramics. Therefore we measured the contact angle in the loaded pole area, as well as in the less stressed inferior area of the femoral head near to the equator. The analysis demonstrated significant higher contact angles in the pole area in comparison to the equator area for both surface modifications, whereas the difference was clearly higher for the LFIT femoral heads. In contrast, alumina femoral heads showed increased contact angles in the equator area. Basically, lower contact angles at the bearing surface should indicate a more hydrophilic surface behavior, resulting in a better wettability [31]. However, within the present simulator study no correlation with contact angles and polyethylene wear could be demonstrated. This was consistent with the findings of Galvin *et al.* [28].

The evaluation of ion levels in the used bovine serum showed that cobalt ion release was higher for the LFIT compared with the TiNbN-coated CoCr heads. Both surface modifications were not able to avoid the ion release. However, with the TiNbN coating the release could be reduced by orders of magnitude. Nevertheless, the ion level found with alumina ceramics ranged around the analytical detection limit of the measuring device, and therefore can be considered for reference.

The investigated surface-modified CoCr femoral heads provide an alternative to ceramic heads for total hip replacement. Within this experimental study, idealized load conditions according to the ISO standard [20] were considered which did not represent all aspects of everyday life activities [32]. Further experimental studies should analyze the effect of adverse conditions and an increased number of load cycles on the wear resistance of coated and ion-treated femoral heads.

5. Conclusions

The wear performance of sequentially cross-linked polyethylene inserts may be improved by ion-treated CoCr and TiNbN-coated CoCr femoral heads. Differences in polyethylene wear were not statistically significant compared with alumina heads. This comparable behavior could be attributed to the increased hardness of the modified CoCr surfaces, leading to more scratch resistance and long-term smoothness. Both surface modifications showed specific wettability, although a correlation with contact angles and polyethylene wear was not detectable within the study.

Cobalt ion release from the substrate could be reduced efficiently by the use of a TiNbN coating in contrast to CoCr heads treated with nitrogen ions.

Acknowledgments: The authors acknowledge Henry Dempwolf and Mario Jackszis from the Department of Orthopaedics, University Medicine Rostock for the preparation of the contact angle images, and supporting the wear measurements. Further acknowledgment goes to Stryker GmbH & Co. KG, Duisburg, Germany, for supporting this study by providing implant components.

Author Contributions: Christian Fabry designed the study, performed the experiments and wrote the initial manuscript. Carmen Zietz performed the experiments, analyzed the data and did the statistical analysis. Axel Baumann prepared the TiNbN coating and supported the data analysis. Rainer Bader was the principal investigator for this research, and designed the study.

Conflicts of Interest: Christian Fabry and Axel Baumann are employees of DOT GmbH, Rostock, Germany. The other authors declare no conflict of interest.

References

1. Affatato, S.; Bordini, B.; Fagnano, C.; Taddei, P.; Tinti, A.; Toni, A. Effects of the sterilisation method on the wear of UHMWPE acetabular cups tested in a hip joint simulator. *Biomaterials* **2002**, *23*, 1439–1446.
2. Medel, F.J.; Kurtz, S.M.; Hozack, W.J.; Parvizi, J.; Purtill, J.J.; Sharkey, P.F.; MacDonald, D.; Kraay, M.J.; Goldberg, V.; Rimnac, C.M. Gamma inert sterilization: A solution to polyethylene oxidation? *J. Bone Joint Surg. Am.* **2009**, *91A*, 839–849.
3. Galvin, A.; Kang, L.; Tipper, J.; Stone, M.; Ingham, E.; Jin, Z.; Fisher, J. Wear of crosslinked polyethylene under different tribological conditions. *J. Mater. Sci. Mater. Med.* **2006**, *17*, 235–243.
4. Kilgour, A.; Elfick, A. Influence of crosslinked polyethylene structure on wear of joint replacements. *Tribol. Int.* **2009**, *42*, 1582–1594.
5. Kurtz, S.M.; Gawel, H.A.; Patel, J.D. History and systematic review of wear and osteolysis outcomes for first-generation highly crosslinked polyethylene. *Clin. Orthop. Relat. Res.* **2011**, *469*, 2262–2277.
6. Lerf, R.; Zurbrugg, D.; Delfosse, D. Use of vitamin E to protect cross-linked UHMWPE from oxidation. *Biomaterials* **2010**, *31*, 3643–3648.
7. Oral, E.; Muratoglu, O.K. Vitamin E diffused, highly crosslinked UHMWPE: A review. *Int. Orthop.* **2011**, *35*, 215–223.
8. Bracco, P.; Oral, E. Vitamin E-stabilized UHMWPE for total joint implants: A review. *Clin. Orthop. Relat. Res.* **2011**, *469*, 2286–2293.
9. Eberhardt, A.W.; McKee, R.T.; Cuckler, J.M.; Peterson, D.W.; Beck, P.R.; Lemons, J.E. Surface roughness of CoCr and ZrO_2 femoral heads with metal transfer: A retrieval and wear simulator study. *Int. J. Biomater.* **2009**, *2009*, 1–6.
10. Dahl, J.; Soderlund, P.; Nivbrant, B.; Nordsletten, L.; Rohrl, S.M. Less wear with aluminium-oxide heads than cobalt-chrome heads with ultra high molecular weight cemented polyethylene cups: A ten-year follow-up with radiostereometry. *Int. Orthop.* **2012**, *36*, 485–490.
11. Wang, S.J.; Zhang, S.D.; Zhao, Y.C. A comparison of polyethylene wear between cobalt-chrome ball heads and alumina ball heads after total hip arthroplasty: A 10-year follow-up. *J. Orthop. Surg. Res.* **2013**, *8*, 1–4.
12. Urban, J.A.; Garvin, K.L.; Boese, C.K.; Bryson, L.; Pedersen, D.R.; Callaghan, J.J.; Miller, R.K. Ceramic-on-polyethylene bearing surfaces in total hip arthroplasty—Seventeen to twenty-one-year results. *J. Bone Joint Surg. Am.* **2001**, *83A*, 1688–1694.
13. Gotman, I.; Hunter, G.; Gutmanas, E.Y. Wear resistant ceramic films and coatings. In *Comprehensive Biomaterials*; Ducheyne, P., Ed.; Elsevier Science: Amsterdam, The Netherlands, 2011; pp. 127–155.
14. McGrory, B.J.; Ruterbories, J.M.; Pawar, V.D.; Thomas, R.K.; Salehi, A.B. Comparison of surface characteristics of retrieved cobalt-chromium femoral heads with and without ion implantation. *J. Arthroplasty* **2012**, *27*, 109–115.

15. Bader, R.; Bergschmidt, P.; Fritsche, A.; Ansorge, S.; Thomas, P.; Mittelmeier, W. Alternative materials and solutions in total knee arthroplasty for patients with metal allergy. *Orthopäde* **2008**, *37*, 136–142.
16. Harman, M.K.; Banks, S.A.; Hodge, W.A. Wear analysis of a retrieved hip implant with titanium nitride coating. *J. Arthroplasty* **1997**, *12*, 938–945.
17. Raimondi, M.T.; Pietrabissa, R. The *in vivo* wear performance of prosthetic femoral heads with titanium nitride coating. *Biomaterials* **2000**, *21*, 907–913.
18. Ries, M.D.; Pruitt, L. Effect of cross-linking on the microstructure and mechanical properties of utra-high molecular weight polyethylene. *Clin. Orthop. Relat. Res.* **2005**, *440*, 149–156.
19. Zietz, C.; Fabry, C.; Middelborg, L.; Fulda, G.; Mittelmeier, W.; Bader, R. Wear testing and particle characterisation of sequentially crosslinked polyethylene acetabular liners using different femoral head sizes. *J. Mater. Sci. Mater. Med.* **2013**, *24*, 2057–2065.
20. ISO 14242-1:2014. *Implants for Surgery—Wear of Total Hip-Joint Prostheses—Part 1: Loading and Displacement Parameters for Wear-Testing Machines and Corresponding Environmental Conditions for Test*; ISO: Genève, Suisse, 2014.
21. Clohisy, J.C.; Calvert, G.; Tull, F.; McDonald, D.; Maloney, W.J. Reasons for revision hip surgery—A retrospective review. *Clin. Orthop. Relat. Res.* **2004**, *429*, 188–192.
22. Ingham, E.; Fisher, J. The role of macrophages in osteolysis of total joint replacement. *Biomaterials* **2005**, *26*, 1271–1286.
23. Abu-Amer, Y.; Darwech, I.; Clohisy, J.C. Aseptic loosening of total joint replacements: Mechanisms underlying osteolysis and potential therapies. *Arthritis Res. Ther.* **2007**, *9*, 1–7.
24. Fisher, J.; Jennings, L.M.; Galvin, A. Wear of highly crosslinked polyethylene against cobalt chrome and ceramic femoral heads. In Bioceramics and Alternative Bearings in Joint Arthroplasty: 11th BIOLOX Symposium Proceedings; Benazzo, F., Falez, F., Dietrich, M., Eds.; Steinkopff Verlag: Darmstadt, Germany, 2006; pp. 185–188.
25. Galvin, A.L.; Tipper, I.L.; Jennings, L.M.; Stone, M.H.; Jin, Z.M.; Ingham, E.; Fisher, J. Wear and biological activity of highly crosslinked polyethylene in the hip under low serum protein concentrations. *Proc. Inst. Mech. Eng. H* **2007**, *221*, 1–10.
26. Galvin, A.L.; Jennings, L.M.; Tipper, J.L.; Ingham, E.; Fisher, J. Wear and creep of highly crosslinked polyethylene against cobalt chrome and ceramic femoral heads. *Proc. Inst. Mech. Eng. H* **2010**, *224*, 1175–1183.
27. Bowsher, J.G.; Williams, P.A.; Clarke, I.C.; Green, D.D.; Donaldson, T.K. "Severe" wear challenge to 36 mm mechanically enhanced highly crosslinked polyethylene hip liners. *J. Biomed. Mater. Res. B Appl. Biomater.* **2008**, *86B*, 253–263.
28. Galvin, A.; Brockett, C.; Williams, S.; Hatto, P.; Burton, A.; Isaac, G.; Stone, M.; Ingham, E.; Fisher, J. Comparison of wear of ultra-high molecular weight polyethylene acetabular cups against surface-engineered femoral heads. *Proc. Inst. Mech. Eng. H* **2008**, *222*, 1073–1080.
29. Gutmanas, E.Y.; Gotman, I. PIRAC Ti nitride coated Ti-6Al-4V head against UHMWPE acetabular cup-hip wear simulator study. *J. Mater. Sci. Mater. Med.* **2004**, *15*, 327–330.
30. Xin, Y.C.; Liu, C.L.; Huo, K.F.; Tang, G.Y.; Tian, X.B.; Chu, P.K. Corrosion behavior of ZrN/Zr coated biomedical AZ91 magnesium alloy. *Surface Coat. Technol.* **2009**, *203*, 2554–2557.
31. Yuan, Y.; Lee, T.R. Contact Angle and Wetting Properties. In *Surface Science Techniques*; Bracco, G., Holst, B., Eds.; Springer: Heidelberg, Germany, 2013; pp. 3–34.
32. Fabry, C.; Herrmann, S.; Kaehler, M.; Klinkenberg, E.D.; Woernle, C.; Bader, R. Generation of physiological parameter sets for hip joint motions and loads during daily life activities for application in wear simulators of the artificial hip joint. *Med. Eng. Phys.* **2013**, *35*, 131–139.

Article

Wear Tests of a Potential Biolubricant for Orthopedic Biopolymers

Martin Thompson [1], Ben Hunt [1], Alan Smith [2] and Thomas Joyce [1,*]

[1] School of Mechanical and Systems Engineering, Newcastle University, Claremont Road, Newcastle upon Tyne NE1 7RU, UK; M.Thompson2@newcastle.ac.uk (M.T.); B.J.Hunt1@newcastle.ac.uk (B.H.)

[2] School of Applied Sciences, University of Huddersfield, Queensgate, Huddersfield HD1 3DH, UK; a.m.smith@hud.ac.uk

* Author to whom correspondence should be addressed; Thomas.joyce@ncl.ac.uk; Tel.: +44-191-208-6214; Fax: +44-191-222-8600.

Academic Editors: Amir Kamali and J. Philippe Kretzer
Received: 16 January 2015; Accepted: 5 March 2015; Published: 25 March 2015

Abstract: Most wear testing of orthopedic implant materials is undertaken with dilute bovine serum used as the lubricant. However, dilute bovine serum is different to the synovial fluid in which natural and artificial joints must operate. As part of a search for a lubricant which more closely resembles synovial fluid, a lubricant based on a mixture of sodium alginate and gellan gum, and which aimed to match the rheology of synovial fluid, was produced. It was employed in a wear test of ultra high molecular weight polyethylene pins rubbing against a metallic counterface. The test rig applied multidirectional motion to the test pins and had previously been shown to reproduce clinically relevant wear factors for ultra high molecular weight polyethylene. After 2.4 million cycles (125 km) of sliding in the presence of the new lubricant, a mean wear factor of 0.099×10^{-6} mm^3/Nm was measured for the ultra high molecular weight polyethylene pins. This was over an order of magnitude less than when bovine serum was used as a lubricant. In addition, there was evidence of a transfer film on the test plates. Such transfer films are not seen clinically. The search for a lubricant more closely matching synovial fluid continues.

Keywords: biolubricant; ultra high molecular weight polyethylene; wear testing; pin-on-plate; orthopedic; alginate; gellan gum

---

1. Introduction

Total joint replacement is a relatively common and generally very successful procedure. Data from the largest joint registry in the world, the National Joint Registry for England, Wales and Northern Ireland, reports that in the last year for which data is available, 2012–13, over 80,000 hip prostheses and 85,000 knee prostheses were implanted in these countries [1]. The registry also states that, at 10 years follow up, the revision rate for cemented hips and cemented knees was only 3.20% and 3.33% respectively, thus indicating the long term success of the vast preponderance of these implants [1]. The majority of these hip and knee prostheses consist of a hard metal or ceramic component which articulates against a polyethylene counterface. However, a small number of these implants do need to be revised and in the majority of cases this is due to wear induced osteolysis [2,3]. Here the polyethylene wear debris provokes a negative cascade of events within the body eventually leading to osteolysis and a revision operation. Therefore the issue of wear is a long-term problem in joint replacements.

As such it is essential both to understand and to minimize the wear processes taking place, and tribological studies have been widely undertaken to study the wear of polyethylene and other orthopedic biopolymers *in vitro*. A key element in such testing has been the appropriate choice of lubricant [4].

Dilute bovine serum is currently recommended as the lubricant for wear testing orthopedic biopolymers [5–7]. This is because: it results in clinically relevant wear rates; it prevents the formation of a transfer film (and such transfer films are not seen on explanted joints); and it results in wear debris which matches the size and shape of polyethylene wear debris seen *in vivo* [4,8]. However there are recognized issues with this lubricant including batch to batch variation, cost and safety [4]. As a biological material it also needs to be changed regularly and this will likely remove wear particles which can influence the tribological performance. For these and other reasons, comparing wear results between different labs can be problematic. Moreover, while it can be fascinating from a tribological view to investigate the constituents of bovine serum and their effect on wear performance, it must also be accepted that bovine serum lacks key elements which exist within synovial fluid and are known to influence the tribology of joints. Likewise it should be self-evident that it is not bovine serum but synovial fluid in which artificial joints must operate [4]. The ideal would be to have a biolubricant which is safe, relatively inexpensive, mimics the properties of synovial fluid and which does not need to be replaced at frequent intervals. The current paper is one contribution towards this overall ideal.

For all of these reasons alternative lubricants have been sought and tested [9]. To add to this body of data a new lubricant was investigated which has been shown to mimic certain rheological properties of synovial fluid [10]. Wear tests were undertaken in a screening wear tester which had previously been shown to produce clinically relevant wear factors for orthopedic biopolymers [11–17]. Details of the new lubricant, alongside comparable properties of bovine serum and human synovial fluid, are given in Table 1. It should be noted that the characteristics of synovial fluid, as a biological fluid, will cover a spectrum and will vary depending on the individual as well as any arthritic disease that may be present [4].

Table 1. Comparative table of lubricant properties. * Data taken from [4]. † Data taken from [10].

	Bovine Serum	Human Synovial Fluid	New Lubricant
Protein	Yes (60 g/L) *	Yes (17 g/L) *	No
Polysaccharide	None *	Hyaluronic acid (3.2 g/L) *	Sodium alginate 2% *w/w* and Gellan gum 0.75% *w/w*
Phospholipids	None *	Yes (0.13–1.15 g/L) *	None †
Viscosity across Shear rates 0.1–10 (s^{-1})	0.1–0.005 Pas †	5–0.05 Pas †	1–0.05 Pas †
Elastic Modulus (at 1 rad^{-1})	~0.01 Pa †	~0.5 Pa †	~0.75 Pa †

2. Results and Discussion

A polysaccharide solution consisting of a 50:50 mix of 2% *w/w* alginate and 0.75% *w/w* gellan gum was prepared and investigated as a lubricant for wear testing orthopedic implant materials. This lubricant was shown to have the non-Newtonian characteristics similar to that of aspirated synovial fluid with a reduction in dynamic viscosity with increasing shear rate (Figure 1). Furthermore, the viscosity of both the synovial fluid and the new lubricant was a factor of 10 greater than the bovine serum across all the shear rates measured.

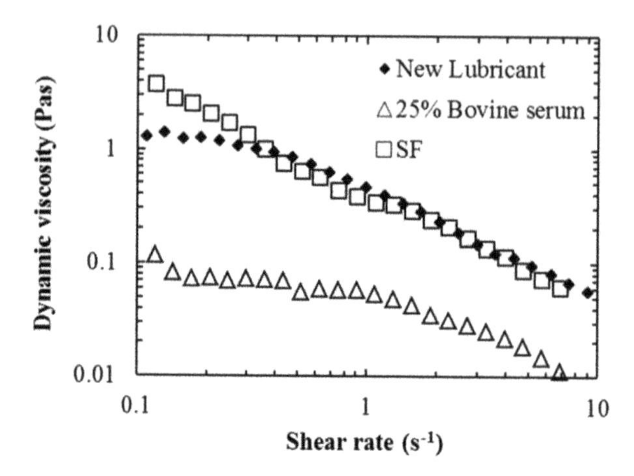

Figure 1. Dynamic viscosity *vs.* sheer rate for: aspirated synovial fluid (SF) (open squares); 25% *w/v* bovine serum (open triangles); and the new lubricant (50:50 mix of 2% *w/w* alginate and 0.75% *w/w* gellan gum) (filled diamonds); measurements undertaken at 37 °C. Data adapted from [10].

The rheological disparity of bovine serum, and similarity of the new lubricant to that of synovial fluid, is further highlighted in Figure 2. Synovial fluid has a mechanical spectra characteristic of a concentrated entangled biopolymer solution meaning the storage modulus (G′) and loss modulus (G″) (G″ > G′ at low frequencies of oscillation and G′ > G″ at high frequencies) are mimicked by the new lubricant. Moreover, the moduli measured in the synovial fluid and the new lubricant samples were over an order of magnitude greater than that of the bovine serum. It is thought that the viscoelastic behavior of the alginate/gellan mixture is due to the alginate providing the viscous response at low frequencies and the gellan contributing to the elastic response at high frequencies [10].

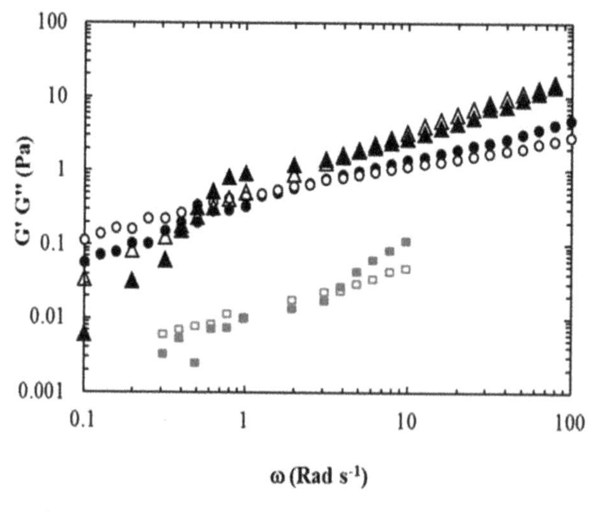

Figure 2. Mechanical spectra (2% strain; 37 °C) for: synovial fluid (G′ filled circles and G″ open circles); the new lubricant (G′ filled triangles and G″ open triangles); and dilute bovine serum 25 g/L protein (G′ filled squares and G″ open squares). Data adapted from [10].

After 2.4 million cycles (125 km) of sliding, the mean volumetric wear rate of the ultra high molecular weight polyethylene (UHMWPE) test pins were 0.45 mm^3/million cycles. This was equivalent to a mean wear factor of 0.099 $\times 10^{-6}$ mm^3/Nm. Weight changes were measured for each pin at 12 intervals during the 125 km of testing. Wear factors for each test pin, corrected for the control pins, are shown in Table 2. The control pins increased in weight, and this increase fluctuated in magnitude over the duration of testing, but at the end of testing an increase of 110 µg was measured. In comparison, at the end of testing, the four test pins showed a mean weight loss of 115 µg. Plate surface roughness values changed from a mean of 0.015 µm Rq, prior to the test, to 0.029 µm Rq at the end of testing; as shown in Table 3. Rq is the root mean square roughness. No noticeable changes in the characteristics of the new testing fluid over the duration of the test were observed.

Table 2. Mean wear factors of the four ultra high molecular weight polyethylene (UHMWPE) test pins after 125 km of sliding; also the final roughness values.

Pin No.	Wear Factor (*k*) ($\times 10^{-6}$ mm^3/Nm)	Standard Deviation ($\times 10^{-6}$ mm^3/Nm)	Mean Rq after Test
1	0.120		2.188 µm
2	0.133		1.303 µm
3	0.078		2.279 µm
4	0.063		2.430 µm
Average	0.099	0.034	2.050 µm

Table 3. Mean roughness values of the wear tracks of the four test plates before testing and at the end of testing.

Plate No.	Mean Rq before Test	Mean Rq after Test
1	0.012 µm	0.019 µm
2	0.015 µm	0.036 µm
3	0.018 µm	0.031 µm
4	0.014 µm	0.027 µm
Average	0.015 µm	0.028 µm

When the same biomaterials were tested in the same rig using a lubricant of dilute bovine serum a mean wear factor of 1.6 $\times 10^{-6}$ mm^3/Nm was measured [18]. This is close to the reported mean wear factor of 2.1 $\times 10^{-6}$ mm^3/Nm for explanted Charnley hips which also used the same biomaterials of stainless steel and UHMWPE [19]. With the new lubricant, the average mean wear factor was 0.099 $\times$ 10^{-6} mm^3/Nm and therefore over an order of magnitude lower than with dilute bovine serum.

The plate surface roughness values at the end of the test were higher than at the beginning of the test and this may indicate the presence of a transfer film. From the non-contacting profilometer measurements, the key feature was multi-directional scratching (Figure 3). In addition surface adhesions were seen on the wear tracks of the test plates (Figure 3) and these adhesions could have originated from the polyethylene pins. Such a transfer film is formed when a hard material, such as a metal, moves against a softer material, such as a polymer, and shears off and picks up a coating of polymer [20]. If the transfer film is stable, then wear rates may be reduced after an initial high wear interval during film formation [20]. Previous work with bovine serum as a lubricant for wear testing UHMWPE pins against a metal counterface has indicated no change in roughness of the metal counterface at the end of testing, at a minimum of 2.5 million cycles, and no transfer film [21,22]. In addition, transfer films of UHMWPE are not seen clinically with such implants [23]. Previously it has been shown that the addition of hyaluronic acid to serum to increase its viscosity had little effect on wear of UHMWPE [18,24]. It may be that, as the sliding velocity is relatively low, so viscosity is not the principal issue in the wear of UHMWPE. Instead, the action of animal-based proteins in boundary lubrication seems to be of high importance as, when animal-based proteins are absent, a transfer film occurs. This has been known for some time with lubricants of distilled water and Ringer solution [25,26] but has also been shown to occur when other novel lubricants (DPPC (dipalmitoylphosphatidylcholine) and soy protein) have been used [27]. A more recent study which wear tested UHMWPE pins against CoCr plates in the presence of 13 different lubricants [9] found that only an egg white based lubricant gave wear factors which were statistically similar to those given by dilute bovine serum. It has been argued for some time that bovine serum serves as a boundary lubricant to prevent a transfer film being formed [28]. In turn it is felt that the proteins within bovine serum allow boundary lubrication without transfer film [27]. As shown by our results, polysaccharides are unable to facilitate boundary lubrication in this application.

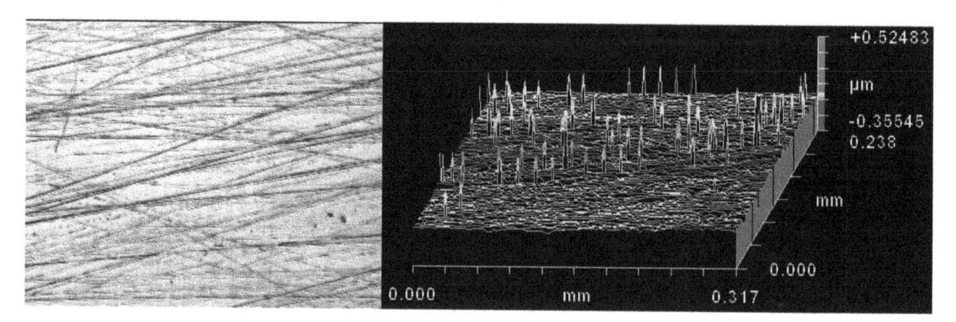

Figure 3. Left hand side image shows an optical image of the worn plate; note the multi-directional scratches; **Right** hand side image shows the equivalent "oblique plot" produced by the ZYGO non-contacting profilometer. Note the peaks which indicate attached material; note too that the horizontal scale is over one thousand times larger than the vertical so that the peaks are not as "severe" as they appear.

For the UHMWPE pins, the mean pre-test roughness was 2.143 μm Rq. While the Rq values at the end of the test were numerically similar, it was noted that the initial concentric machining marks on the pins had largely been removed.

Analysis of the wear debris in the new lubricant revealed particle sizes ranging from ~50 to 400 nm (Figure 4). These nanoparticles are of a similar size range to wear debris found in failed total knee arthroplasties (low contact stress mobile bearing prostheses) [29].

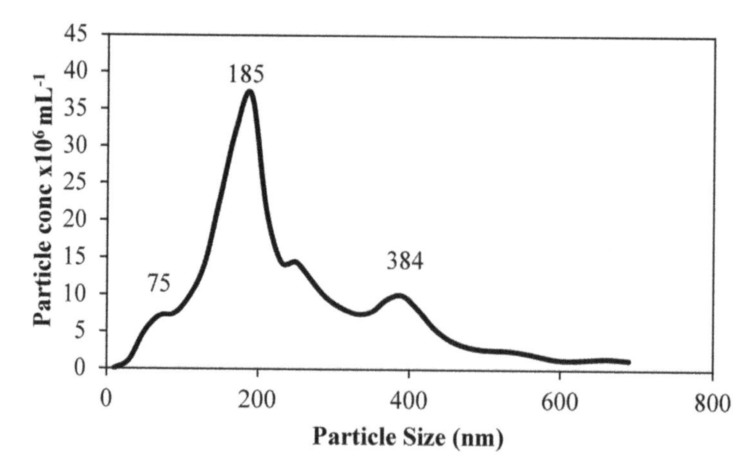

Figure 4. Particle size distribution of wear debris in the new lubricant post test (50:50 mix of 2% *w*/*w* alginate and 0.75% *w*/*w* gellan gum).

There were a number of limitations. Given that the influence of the controls was so important on the overall wear, we could perhaps have employed control pins which were subject to the same axial load as our test pins. However we would point out that: we employed three control pins to try and minimize the effect of lubricant uptake on the overall wear values; unloaded control pins allowed a direct comparison with our previous work which is compared to in the text [18]; and also that it is usual to employ unloaded control pins in such screening wear tests [30–32]. For future work we will look to employing statically loaded control pins. Another limitation was the small test sample size, however such a sample size was in line with previous work [18,31] and the sample size was sufficient to indicate that the new lubricant was unable to match wear factors associated with the use of bovine serum as a lubricant.

3. Experimental Section

The new lubricant consisted of a mixture of sodium alginate and gellan gum, and aimed to match the rheology of synovial fluid. Sodium alginate was used as a synthetic substitute to hyaluronic acid, giving the lubricant non-Newtonian characteristics as seen with synovial fluid, while the gellan gum replaced the lubricin in synovial fluid and aimed to reproduce the viscoelasticity of synovial fluid. Stock solutions of the test lubricants were prepared as previously described [10]. A 50:50 mix of 2% *w*/*w* alginate (Protanal LF200) and 0.75% *w*/*w* gellan gum (kelcogel CG-LA) were subjected to viscosity measurements at 37 °C using a sheer ramp from 0.1–10 s^{-1}. These parameters were chosen to match similar shear rates the lubricant was subjected to during wear testing. Oscillatory shear measurements of storage modulus (G′) and loss modulus (G″) were taken at a constant strain of 2% (within the linear viscoelastic region) across a frequency range of 0.1–100 rad·s^{-1} at 37 °C. Both viscosity measurements and oscillatory measurements were performed on a Bohlin Gemini nano rheometer using a 55 mm parallel plate geometry with a 100 mm gap. All rheological measurements were performed using the set up and parameters as used previously by Smith *et al.* [10].

The four-station wear test rig has been described previously [18]. A schematic image of the rig is offered in Figure 5. As can be seen the key elements are a motor which provided the reciprocating motion to the test bed upon which were held the four test plates. Each test plate sat within an individual bath. Each test pin, which was held within a pin holder, was also subject to a rotational motion, which was provided by a 12 V motor via a pair of spur gears. Each test pin was subject to load which was applied by a weight mounted towards the end of a lever arm.

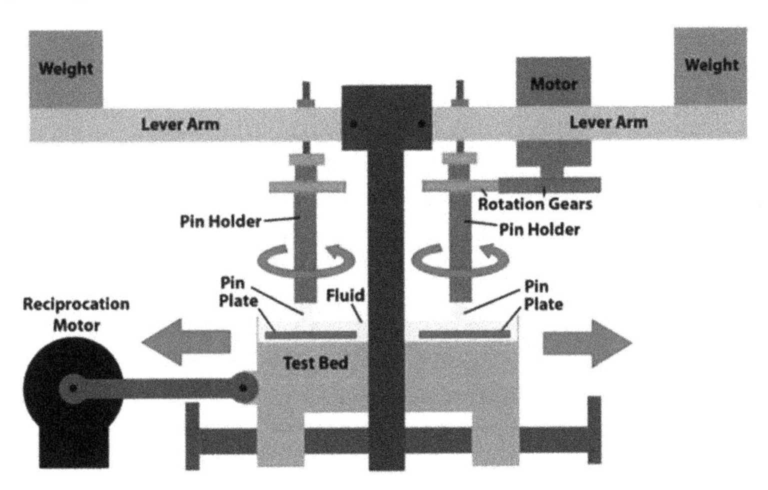

Figure 5. Schematic drawing of the pin-on-plate test rig.

Each of the four test stations applied rotational motion at 1 Hz to 6 mm diameter test pins which were loaded at 40 N against 316 L stainless steel test plates (50 mm × 25 mm × 3 mm) which had been polished to a mean surface finish of 0.015 μm Rq. The 40 N load resulted in a nominal stress of approximately 1.4 MPa. This not only matched that used in previous tests [18] but also fitted well with research which indicates that the average contact pressure in an artificial hip joint is likely within the range 1–2 MPa [22]. A reciprocating motion, again at 1 Hz, was applied to the test plates. The stroke length was 30 mm. Pins were manufactured from UHMWPE and the test pins were subject to multi-directional motion through the combination of the rotational and reciprocating motion. Such a combined motion resulted in each point on the wear face of the test pins following elliptical or quasi-elliptical wear tracks [11], similar to those motions seen on implanted hip prostheses [33]. The lubricant was not heated as it is recognized that, with higher temperatures, increased protein precipitation occurs during testing of hip implants and that this served to decrease wear, through the formation of an adherent layer on the surfaces of artificial hips [34]. Similarly other research has shown that not heating the bulk lubricant to 37 °C resulted in less evaporation of lubricant (so that experimental conditions remained largely unchanged); reduced microbial growth (and thus no need for additives which are both toxic and may change the wear mechanisms); reduced protein precipitation; and, most importantly, wear results which were similar to clinical findings [22].

At regular intervals of approximately 60 h the test was stopped, "test" and "control" lubricant was collected into individual containers, pins and plates were cleaned and weighed three times to a consistent protocol on a balance with a sensitivity of 10 μg. "Control" pins were employed to take account of any lubricant uptake or fluctuations in weight. They were kept in the same test lubricant and subject to cleaning and weighing at the same intervals as the test pins. From such compensated weight changes, a volume change was obtained by using the density of UHMWPE, which was taken to be 949 kg/m^3. Using linear regression and plotting compensated mass loss against sliding distance, the wear rate was computed. Then the wear rate was divided by the density, load and sliding distance to give a wear factor. Thus the wear factor (k, units mm^3/Nm) for each pin was defined as the volume lost (V, units mm^3) divided by the product of the load (L, units N) and the sliding distance (D, units m):

$$k = \frac{V}{LD} \tag{1}$$

Prior to, and at the end of testing, fifty readings of the roughness of the wear track on each of the test plates was measured using a ZYGO 5000 non-contacting profilometer, which had a vertical resolution of better than 1 nm [35,36].

Accumulation of wear debris in the test lubricants was verified using nanoparticle tracking analysis (Nanosight LM10). In order to analyze the wear debris the polysaccharides were removed from the test lubricant by ethanol extraction. Briefly, at the end point of the wear test the lubricant was collected and a diluted 1:100 with ultrapure water (18.2 MΩ·cm). A threefold volume of cold ethanol (95% v/v) was then added and the precipitated polysaccharides were removed with a spatula. The remaining solution was filtered using a Buchner funnel with a pore size of 11 μm. The filtrate was collected and the ethanol removed using a rotary evaporator. The remaining wear debris was then re-suspended in ultrapure water prior to analysis.

4. Conclusions

As currently constituted, the novel lubricant does not reproduce the clinical wear factors associated with failed and explanted metal-on-polyethylene hips. Nor those measured when dilute bovine serum is used as the lubricant for the wear testing of UHMWPE against a metallic counterface. This inconsistency may indicate that a protein component, as is inherent with bovine serum, is essential in a lubricant for wear testing orthopedic biopolymers. This will be an area of future research.

Acknowledgments: No direct research funding for any of the work outlined in this paper was received.

Author Contributions: Thomas Joyce conceived and designed the experiments; Martin Thompson performed the experiments; Ben Hunt undertook topographical measurements; Thomas Joyce and Martin Thompson analyzed the data; Alan Smith contributed the lubricant and related measurements; Thomas Joyce wrote the paper.

Conflicts of Interest: The authors declare no conflict of interest.

References

1. Young, E. *National Joint Registry for England, Wales and Northern Ireland*; 11th Annual Report; National Joint Registry: Hemel Hempstead, UK, 2014.
2. Kim, Y.H.; Park, J.W.; Patel, C.; Kim, D.Y. Polyethylene wear and osteolysis after cementless total hip arthroplasty with alumina-on-highly cross-linked polyethylene bearings in patients younger than thirty years of age. *J. Bone Jt. Surg.* **2013**, *95*, 1088–1093. [CrossRef]
3. Gallo, J.; Goodman, S.B.; Konttinen, Y.T.; Wimmer, M.A.; Holinka, M. Osteolysis around total knee arthroplasty: A review of pathogenetic mechanisms. *Acta Biomater.* **2013**, *9*, 8046–8058. [CrossRef] [PubMed]
4. Harsha, A.P.; Joyce, T.J. Challenges associated with using bovine serum in wear testing orthopaedic biopolymers. *Proc. Inst. Mech. Eng. Part H: J. Eng. Med.* **2011**, *225*, 948–958. [CrossRef]
5. Implants for Surgery—Wear of Total Hip Joint Prostheses, Parts 1 and 2. ISO14242; ISO: Geneva, Switzerland, 2000.
6. Implants for Surgery—Wear of Total Knee Joint Prostheses, Parts 1 and 2. ISO14243; ISO: Geneva, Switzerland, 2009.
7. Standard Test Method for Wear Testing of Polymeric Materials Used in Total Joint Prostheses. ASTM-F732-00; ASTM: West Conshohocken, PA, USA, 2000.
8. Joyce, T.J. Biopolymer tribology. In *Polymer Tribology*; Sinha, S.K., Briscoe, B.J., Eds.; Imperial College Press: London, UK, 2009; pp. 227-266.
9. Scholes, S.C.; Joyce, T.J. *In vitro* tests of substitute lubricants for wear testing orthopaedic biomaterials. *Proc. Inst. Mech. Eng. Part H: J. Eng. Med.* **2013**, *227*, 693–703. [CrossRef]
10. Smith, A.M.; Fleming, L.; Wudebwe, U.; Bowen, J.; Grover, L.M. Development of a synovial fluid analogue with bio-relevant rheology for wear testing of orthopaedic implants. *J. Mech. Behav. Biomed. Mater.* **2014**, *32*, 177–184. [CrossRef] [PubMed]
11. Joyce, T.J.; Unsworth, A. A multi-directional wear screening device and preliminary results of UHMWPE articulating against stainless steel. *Bio-Med. Mater. Eng.* **2000**, *10*, 241–249.
12. Joyce, T.J. Biopolymer wear screening rig validated to astm f732-00 and against clinical data. *Tribol. Mater. Surf. Interfaces* **2007**, *1*, 63–67. [CrossRef]
13. Joyce, T.J.; Unsworth, A. A comparison of the wear of cross-linked polyethylene against itself with the wear of ultra-high molecular weight polyethylene against itself. *J. Eng. Med.* **1996**, *210*, 297–300. [CrossRef]
14. Joyce, T.J.; Vandelli, C.; Cartwright, T.; Unsworth, A. A comparison of the wear of cross-linked polyethylene against itself under reciprocating and multi-directional motion with different lubricants. *Wear* **2001**, *250*, 206–211. [CrossRef]
15. Joyce, T.J.; Thompson, P.; Unsworth, A. The wear of ptfe against stainless steel in a multi-directional pin-on-plate wear device. *Wear* **2003**, *255*, 1030–1033. [CrossRef]
16. Joyce, T.J.; Unsworth, A. Wear studies of all UHMWPE couples under various bio-tribological conditions. *J. Appl. Biomater. Biomech.* **2004**, *2*, 29–34. [PubMed]
17. Joyce, T.J. The wear of two orthopaedic biopolymers against each other. *J. Appl. Biomater. Biomech.* **2005**, *3*, 141–146. [PubMed]
18. Joyce, T.J. Wear tests of orthopaedic biopolymers with the biolubricant augmented by a visco-supplement. *Proc. Inst. Mech. Eng. Part J: J. Eng. Tribol.* **2009**, *223*, 297–302. [CrossRef]
19. Hall, R.M.; Unsworth, A.; Siney, P.D.; Wroblewski, B.M. Wear in retrieved charnley acetabular sockets. *J. Eng. Med.* **1996**, *210*, 197–207. [CrossRef]
20. Black, J. *Biological Performance of Materials: Fundamentals of Biocompatibility*; CRC Press: Boca Raton, FL, USA, 2006.
21. Harsha, A.P.; Joyce, T.J. Comparative wear tests of ultra-high molecular weight polyethylene and cross-linked polyethylene. *Proc. Inst. Mech. Eng. Part H: J. Eng. Med.* **2013**, *227*, 600–608. [CrossRef]

22. Saikko, V. A hip wear simulator with 100 test stations. *Proc. Inst. Mech. Eng. Part H: J. Eng. Med.* **2005**, *219*, 309–318. [CrossRef]
23. Wang, A.; Sun, D.C.; Stark, C.; Dumbleton, J.H. Wear mechanisms of UHMWPE in total joint replacements. *Wear* **1995**, *181–183*, 241–249. [CrossRef]
24. Wang, A.; Essner, A.; Schmidig, G. The effects of lubricant composition on *in vitro* wear testing of polymeric acetabular components. *J. Biomed. Mater. Res. (Appl. Biomater.)* **2004**, *68B*, 45–52. [CrossRef]
25. McKellop, H.A. Wear of artificial joint materials ii. Twelve-channel wear-screening device: Correlation of experimental and clinical results. *Eng. Med.* **1981**, *10*, 123–136. [CrossRef]
26. Kumar, P.; Oka, M.; Ikeuchi, K.; Shimizu, K.; Yamamuro, T.; Okumura, H.; Koloura, Y. Low wear rates of UHMWPE against zirconia ceramic in comparison to alumina ceramic and sus 316l alloy. *J. Biomed. Mater. Res.* **1991**, *25*, 813–828. [CrossRef] [PubMed]
27. Saikko, V.; Ahlroos, T. Type of motion and lubricant in wear simulation of polyethylene acetabular cup. *J. Eng. Med.* **1999**, *213*, 301–310. [CrossRef]
28. Wang, A.; Essner, A.; Stark, C.; Dumbleton, J.H. Comparison of the size and morphology of UHMWPE wear debris produced by a hip joint simulator under serum and water lubricated conditions. *Biomaterials* **1996**, *17*, 865–871. [CrossRef] [PubMed]
29. Huang, C.H.; Ho, F.Y.; Ma, H.M.; Yang, C.T.; Liau, J.J.; Kao, H.C.; Young, T.H.; Cheng, C.K. Particle size and morphology of UHMWPE wear debris in failed total knee arthroplasties—A comparison between mobile bearing and fixed bearing knees. *J. Orthop. Res.* **2002**, *20*, 1038–1041. [CrossRef] [PubMed]
30. Saikko, V. Effect of lubricant protein concentration on the wear of ultra-high molecular weight polyethylene sliding against a cocr counterface. *Trans. Am. Soc. Mech. Eng. J. Tribol.* **2003**, *125*, 638–642.
31. Bragdon, C.R.; O'Connor, D.O.; Lowenstein, J.D.; Jasty, M.; Biggs, S.A.; Harris, W.H. A new pin-on-disk wear testing method for simulating wear of polyethylene on cobalt-chrome alloy in total hip arthroplasty. *J. Arthroplast.* **2001**, *16*, 658–665. [CrossRef]
32. Lancaster, J.G.; Dowson, D.; Fisher, J.; Isaac, G.H. The wear of ultra-high molecular weight polyethylene sliding on metallic and ceramic counterfaces representative of current femoral surfaces in joint replacement. *Proc. Inst. Mech. Eng. Part H: J. Eng. Med.* **1997**, *211*, 17–24. [CrossRef]
33. Ramamurti, B.S.; Estok, D.M.; Jasty, M.; Harris, W.H. Analysis of the kinematics of different hip simulators used to study wear of candidate materials for the articulation of total hip arthroplasties. *J. Orthop. Res.* **1998**, *16*, 365–369. [CrossRef] [PubMed]
34. Liao, Y.S.; McKellop, H.A.; Lu, Z.; Campbell, P.; Benya, P. The effect of frictional heating and forced cooling on the serum lubricant and wear of uhmw polyethylene cups against cobalt-chromium and zirconia balls. *Biomaterials* **2003**, *24*, 3047–3059. [CrossRef] [PubMed]
35. Joyce, T.J.; Langton, D.J.; Jameson, S.S.; Nargol, A.V.F. Tribological analysis of failed resurfacing hip prostheses and comparison with clinical data. *Proc. Inst. Mech. Eng. Part J: J. Eng. Tribol.* **2009**, *223*, 317–323. [CrossRef]
36. Joyce, T.J.; Grigg, H.; Langton, D.J.; Nargol, A.V.F. Quantification of self-polishing *in vivo* from explanted metal-on-metal total hip replacements. *Tribol. Int.* **2011**, *44*, 513–516. [CrossRef]

 *lubricants*

Article

In Vitro Wear Testing of a CoCr-UHMWPE Finger Prosthesis with Hydroxyapatite Coated CoCr Stems

Andrew Naylor [1,*], Sumedh C. Talwalkar [2], Ian A. Trail [2] and Thomas J. Joyce [1]

1 School of Mechanical and Systems Engineering, Newcastle University, Newcastle upon Tyne, England NE1 7RU, UK; thomas.joyce@newcastle.ac.uk

2 Upper Limb Research Unit, Wrightington Hospital, Wigan, England WN6 9EP, UK; sctalwalkar@gmail.com (S.C.T.); ian.trail@wwl.nhs.uk (I.A.T.)

* Author to whom correspondence should be addressed; andrew.naylor@newcastle.ac.uk; Tel.: +44-0-7580-207-250.

Academic Editors: Amir Kamali and J. Philippe Kretzer
Received: 3 March 2015; Accepted: 2 April 2015; Published: 13 April 2015

Abstract: A finger prosthesis consisting of a Cobalt-chromium (CoCr) proximal component and an Ultra-high-molecular-weight-polyethylene (UHMWPE) medial component (both mounted on hydroxyapatite coated stems) was evaluated to 5,000,000 cycles in an *in vitro* finger simulator. One "test" prosthesis was cycled through flexion-extension (90°–30°) with a dynamic load of 10 N, whilst immersed in a lubricant of dilute bovine serum. Additionally, a static load of 100 N was applied for 45 s every 3000 cycles to simulate a static gripping force. A second "control" prosthesis was immersed in the same lubricant to account for absorption. Gravimetric and Sa (3D roughness) measurements were taken at 1,000,000 cycle intervals. Micrographs and Sa values revealed negligible change to the CoCr surfaces after 5,000,000 cycles. The UHMWPE also exhibited no distinctive Sa trend, however the micrographs indicate that polishing occurred. Both the CoCr and UHMWPE test components progressively decreased in weight. The CoCr control component did not change in weight, whilst the UHMWPE component gained weight through absorption. To account for the disparity between surface and gravimetric results, the hydroxyapatite coatings were examined. Micrographs of the test stems revealed that the hydroxyapatite coating was partially removed, whilst the micrographs of the control stems exhibited a uniform coating.

Keywords: wear analysis/testing; joint simulators; biotribology; biomechanical testing/analysis; surfaces

1. Introduction

The proximal interphalangeal (PIP) is the joint that connects the first and second phalanges of the finger. The two main indications for surgical replacement of the joint (arthroplasty) are osteoarthritis and rheumatoid-arthritis, both of which affect mainly women over the age of 50 [1,2]. The Swanson single piece silicone prosthesis is considered to be the "replacement of choice" for finger joint arthroplasty [3], and has been used extensively [4–8]. It has, however, been argued that surface replacements (SR) with a bicondylar design have the potential to restore greater post-operative function and range of motion [3], since they replicate the natural anatomy of the PIP joint. The prosthesis used in this present study is such a SR design.

The MatOrtho proximal interphalangeal replacement (PIPR), formerly the Finsbury PIPR (Figure 1a), is an anatomical SR prosthesis used for arthroplasty of the PIP joint. The articulating surface materials are Cobalt-Chromium (CoCr) for the proximal component; and Ultra High Molecular Weight Polyethylene (UHMWPE) for the medial component; both of which are mounted on CoCr stems with an external hydroxyapatite coating. A short-term study [9], with a minimum follow up of 12 months, evaluated 43 implanted prostheses and. reported improvements in pain, function and range of motion. However it has been acknowledged that longer-term study is required to assess the longevity of the prosthesis [10].

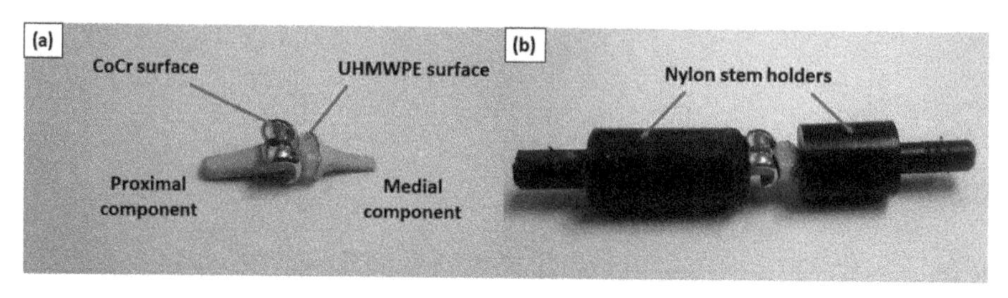

Figure 1. (**a**) The unconstrained prosthesis pair with articulating surfaces labelled and (**b**) the prosthesis pair constrained in the nylon stem holders.

The Small Bone Innovations (formerly Avanta Orthopaedics) SR PIP (surface replacement proximal interphalangeal prosthesis) has yielded mixed results clinically [11–14]. This prosthesis has the same articulating surface materials (CoCr and UHMWPE), but is mounted on porous titanium stems rather than hydroxyapatite coated CoCr stems. Two clinical studies with mean follow-up periods of 15 [12] and 27 months [14] evaluated 67 and 20 PIP prostheses respectively. Both studies found significantly reduced pain and negligible loss in the preoperative range of motion. However, high rates of loosening have also been reported for the SR PIP, with one study reporting a 33% rate of loosening [11] and another reporting a 68% rate [13]. This loosening indicates that the titanium stems have not integrated well with the cancellous bone.

A different type of CoCr-UHMWPE prosthesis (the Kls-Martin CapFlex PIP) was recently evaluated [15], with 10 prostheses observed over 12 months. This study reported no cases of loosening with overall improvements to strength, range of motion and pain relief. The key difference between this prosthesis and the two other mentioned anatomical prostheses (SR PIP and MatOrtho PIPR) is a shorter central stem, with additional short stems either side to improve lateral joint stability.

To date, no *in vitro* wear testing of any of the above mentioned CoCr on UHMWPE prosthesis designs has been reported in the scientific literature, despite being implanted into hundreds of people. The advantage of *in vitro* testing is that throughout the duration of testing, test conditions can be carefully controlled. Testing can also be paused at regular intervals to allow for precise measurements to be taken. With a clinical study, the prostheses are *in situ* limiting the analysis that can be performed. The aim of this present study was to determine the wear characteristics of the CoCr and UHMWPE surfaces through such a program of *in vitro* wear testing.

2. Methods

A MatOrtho PIPR size 8 proximal (CoCr) component with a nominal weight of 1.55 g was tested against a size 7 medial (UHMWPE) component with a nominal weight of 0.45 g (Figure 1a). This combination was tested to five million cycles of flexion-extension, with a 60° arc of motion used to represent the functional "day to day" range of motion in the hand [16]. Other recommended test conditions [17] were followed, including: Limiting the test frequency to 1.5 Hz; and use of dilute bovine serum (maintained at a temperature of 37 °C) as a lubricant. The protein concentration used was 20 g/L, which met the requirements for testing joint prostheses *in vitro* [18,19]. It has been demonstrated in a previous study [20] that CoCr-UHMWPE prostheses adhere to a boundary lubrication regime when tested in dilute bovine serum at low frequency. This was the case for prostheses with equivalent spherical radii ranging from 3 to 10 mm. It is evident that boundary lubrication would not be desirable for prosthesis pairs of the same material (metal-on-metal, ceramic-on-ceramic); however, a previous *in vitro* experiment [21] found that a metal-on-polymer prosthesis had relatively low wear after five million cycles of flexion-extension, despite adhering to an apparent boundary lubrication regime.

The same combination of sizes was used as a "control" prosthesis pair, which did not undergo any flexion-extension cycles but was however immersed in the same lubricant in the test chamber to account for any lubricant uptake. The test regime was identical to that used by Naylor *et al.* for the evaluation of the Ascension Pyrocarbon PIPJ [22,23], and the *in vitro* finger simulator (Figure 2) has been described in much greater detail elsewhere [6,24,25]. A brief description of the test procedure and simulator is provided below.

The flexion-extension mechanism was driven by two small (10 mm diameter) pneumatic cylinders, the pistons of which were connected to braided polyethylene "flexor" and "extensor" tendons which translated through a pulley system to connect to the nylon medial component holder. The proximal component remained stationary throughout testing, with the medial component holder oscillating about the axis of the proximal component. Regulators were used to ensure that the pressure entering each of the dynamic cylinders was 1.3 bar, which in turn was measured using pressure gauges. This resulted in dynamic loads of approximately 10 N in both flexion and extension, which is consistent with dynamic finger activity such as operating a keyboard [26].

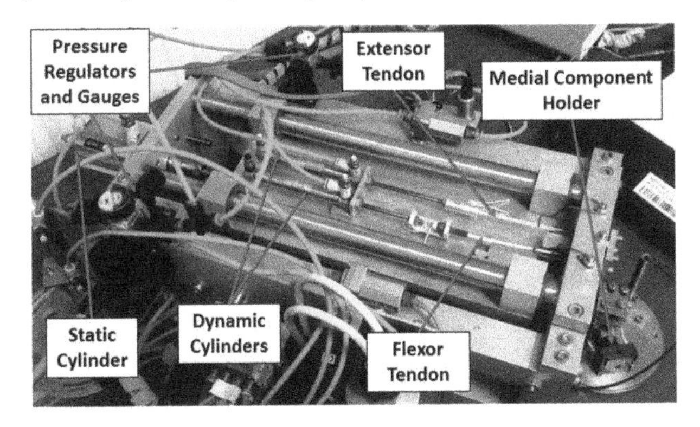

Figure 2. The finger simulator used to conduct the experimentation.

To provide a biomechanical simulation of gripping and pinching, the dynamic cylinders momentarily paused at 3000 cycles to allow for a larger static cylinder (32 mm diameter) to actuate. This cylinder moved the entire platform on which the dynamic cylinders were fixed, subsequently pulling the flexor tendon in tension (air was dumped from the extensor cylinder). As with the dynamic cylinders, a regulator was used to ensure that the pressure entering the "static" cylinder was 1.3 bar, verified by an inline pressure gage. This resulted in a static load of approximately 100 N, which conforms to the strength of the arthritic finger [27,28].

Nylon has been commonly used as an orthopaedic bone substitute [29,30], with a yield stress similar to cortical bone ranging from approximately 50–100 MPa and hence was selected to be an appropriate "bone substitute" material for the stem holders (Figure 1b). A light push fit was incorporated into the design of the stem holders to prevent displacement during flexion-extension. Such a holder and fit allowed the test components to be removed for topographic and gravimetric measurements.

Gravimetric measurements were taken at one million cycle intervals, for a total of five million cycles. First the prosthetic components were removed and sterilized using Virkon disinfectant. The articulating surfaces were then subject to further cleaning using isopropanol and lint free cloths. After allowing approximately half an hour for any residual moisture to evaporate, the prosthetic components were then weighed using a Denver Instruments TB215D balance to a sensitivity of 0.1 mg. The CoCr proximal and UHMWPE medial components were weighed separately to the nearest

0.1 mg to determine any gravimetric change. 3D roughness (Sa) values of the articulating surfaces were obtained using a ZYGO NewView 5000 optical surface profiler with a resolution of 1 nm [31]. For the proximal component five Sa measurements were taken at and around the apex of each of the two CoCr condyles. The geometric constraints of the UHMWPE medial component limited the surface roughness measurements to one at each of the "poles" of the UHMWPE convex surfaces. This resulted in ten Sa measurements for each CoCr surface and a further two values for the UHMWPE surfaces. As with the gravimetric measurements, this process was conducted prior to testing, then at one million cycle intervals during testing. Micrographic analysis of the stems was also conducted using the optical surface profiler to provide a visual comparison of the test and control stems at the end of testing.

3. Results

3.1. Surface Analysis

Prior to testing, the measured Sa values for the CoCr test surfaces ranged from 28 to 57 nm, compared with 34 to 81 nm for the control. At five million test cycles the Sa values for the test surfaces ranged from 35 to 68 nm, compared with 26 to 53 nm for the control. Sa values for the UHMWPE surfaces taken across five million cycles ranged from 1100 to 2900 nm, which although two orders of magnitude higher than the CoCr components, is still representative of roughness values for UHMWPE finger components [20,32]. No distinctive trends emerged from the Sa measurements for either the CoCr or the UHMWPE test components over the five million test cycles (Figure 3). Correspondingly, micrographic images of the CoCr surfaces exhibited little or no change after five million cycles (Figure 4). Micro-machining marks were visible on the surface of the UHMWPE test surfaces prior to testing, but had been removed by five million cycles (Figure 5).

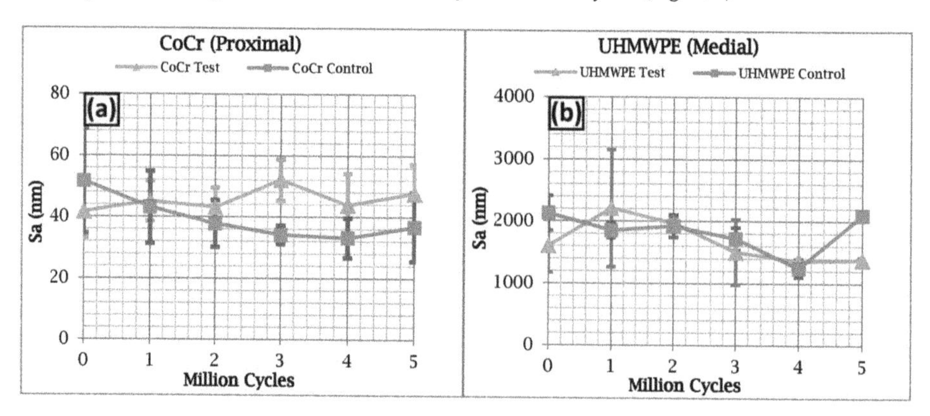

Figure 3. 3D surface roughness (Sa) results for (**a**) the cobalt-chromium (CoCr) proximal surfaces and (**b**) the ultra-high-molecular-weight-polyethylene (UHMWPE) surfaces.

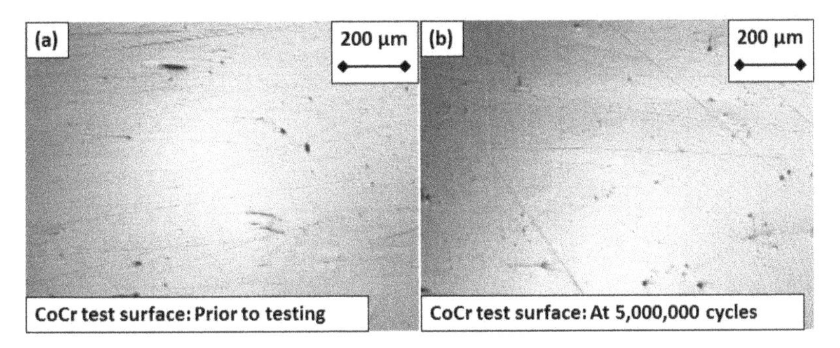

Figure 4. Micrograph images of the CoCr articulating "test" surface (**a**) prior to testing and (**b**) after five million cycles of flexion and extension.

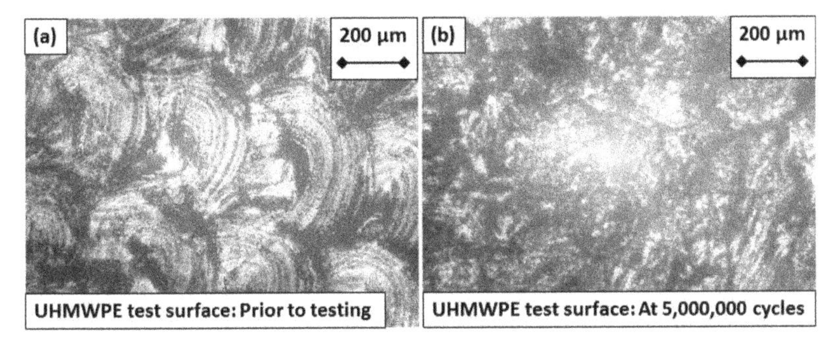

Figure 5. Micrograph images of the UHMWPE articulating "test" surface (**a**) prior to testing and (**b**) after five million cycles of flexion and extension.

3.2. Gravimetric Analysis

In contrast to the surface roughness evaluation, a distinctive trend was exhibited for change in mass (Δm) over the five million cycles for both CoCr proximal and the UHMWPE medial test components (Figure 6a). The apparent rate of gravimetric loss for the proximal CoCr test component was approximately 0.7 mg per million cycles, and 0.3 mg per million cycles for the medial UHMWPE test component. The control CoCr proximal component exhibited a negligible change in mass over the five million test cycles. In comparison the control medial UHMWPE component gained mass through the absorption of bovine serum at a substantial rate of 0.1 mg per million test cycles. The gain in mass of the control components was subtracted from the absolute gravimetric values (Figure 6a) to provide the more realistic relative gravimetric values (Figure 6b). These compensated rates of gravimetric loss were 0.7 mg and 0.4 mg per million cycles for the proximal CoCr and medial UHMWPE test components respectively.

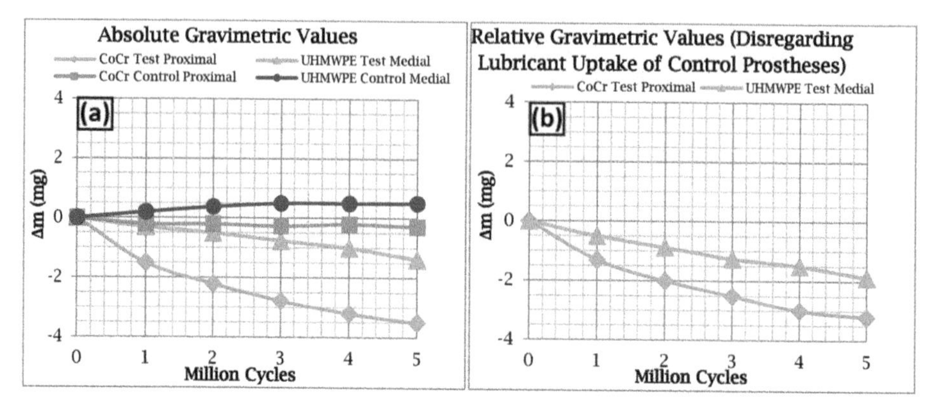

Figure 6. Gravimetric results for (**a**) all of the test and control components; and (**b**) The CoCr and UHMWPE test components minus the "soak" acquired by the control components.

3.3. *Micrographic Analysis of the Stems*

For the control prosthesis, the stems of both the proximal and medial components exhibited a uniform coating of hydroxyapatite at five million cycles of testing (Figure 7). In contrast, for the test prosthesis, the stems of the proximal and medial components exhibited a less uniform coating of hydroxyapatite (Figure 8), showing areas where the coating has been partially removed.

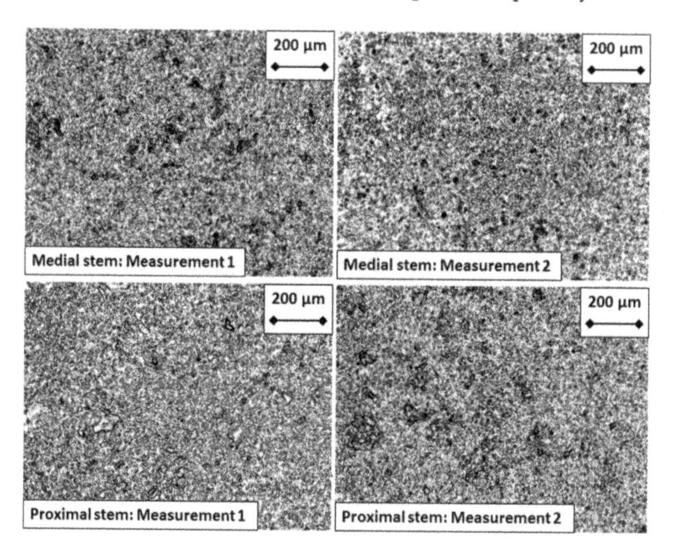

Figure 7. Micrographs of the Hydroxyapatite coating taken from the control prosthesis proximal and medial stems, at the end of experimentation.

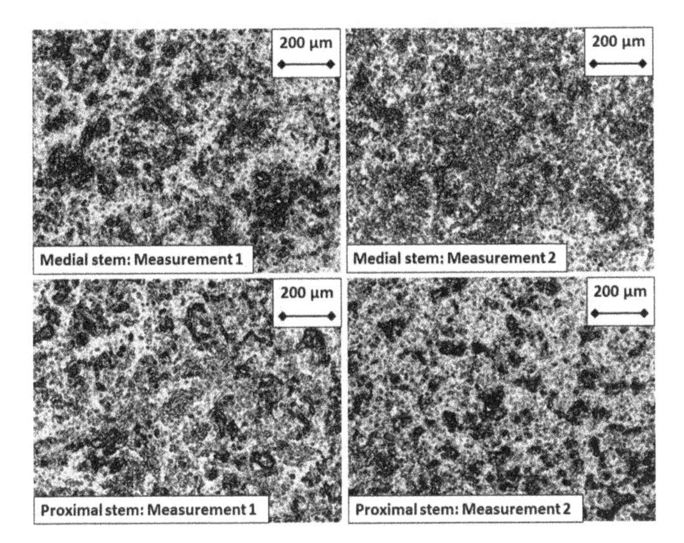

Figure 8. Micrographs of the Hydroxyapatite coating taken from the test prosthesis proximal and medial stems, after five million cycles of flexion and extension.

4. Discussion

No relationship was found between the roughness of the articulating surfaces and the progressive number of test cycles (Figure 3). This was supported by the fact that, visually, the CoCr test surface did not exhibit any signs of wear at five million test cycles (Figure 4). The micro-machining marks visible on the UHMWPE test surface prior to testing gradually faded with each million cycles of testing, yet as with the CoCr surface, neither an increase nor decrease in Sa values were observed over the duration of the experiment. The gradual fading of the micro-machining marks was consistent with a polishing effect.

From a previous *in vitro* study [32], comparative roughness values, taken from the articulating surfaces of a stainless steel (SS) *versus* UHMWPE two-part finger prosthesis (Digital Joint Operative Arthroplasty (DJOA); Landos, Chaumont, France) provided evidence of a polishing effect on the UHMWPE. Initially the UHMWPE surface had a roughness of 1300 nm, which decreased to 270 nm after seven million cycles of flexion-extension testing. Measurements were made with a Mitutoyo Form-tracer 4100, providing roughness in 2D rather than the 3D offered by the ZYGO NewView 5000 optical profiler. The SS surface had an initial roughness of 147 nm, which increased to 209 nm after seven million test cycles. When comparing these results with this present study it must be stated that the CoCr surface was lower in roughness than the SS surface, and further it did not exhibit any change in roughness over the five million test cycles. In part these differences could be explained by the higher hardness of the CoCr compared with the SS, which meant it was more scratch resistant. A potential explanation for the differences in initial roughness values is that the polishing process employed by the manufacturer of the contemporary MatOrtho prosthesis may have been superior to that of the DJOA which dates from the 1980s.

Changes in weight due to lubricant uptake can be an important factor. Previous *in vitro* studies evaluating the polymeric components of two-part finger prostheses [21,32,33], tested over a comparable range of cycles to this study, have reported an increase in weight due to lubricant uptake. This uptake has often been for both the test and the control polymeric components [21,32,33]. UHMWPE [21,32] test components increased in weight, but the gravimetric measurements for the control components showed a similar increase to that of the test components. Interestingly, cross linked polyethylene (XLPE) [33] test components increased in weight, but no more than those of the control. For both materials, UHMWPE and XLPE, the results indicated that no measurable wear occurred. In contrast to this, a poly ether ether ketone (PEEK) [21] test component from a finger prosthesis increased in weight, but less than the PEEK control component. This difference indicated that wear of the PEEK component occurred. The UHMWPE test component evaluated in this present study decreased in weight, whilst the UHMWPE control component increased. This is not consistent with the previous *in vitro* studies and indicates that a different "wear" mechanism occurred for the UHMWPE medial component. Gravimetric results for a Stainless Steel (SS) [32] proximal component tested against an UHMWPE component showed no observable change for either the SS test or SS control components. Although this is comparable to the "control" CoCr component from this present study, which also exhibited no gravimetric change, it is not comparable with the "test" CoCr component, which exhibited gravimetric loss over five million test cycles. Once again, this indicated that a different "wear" mechanism occurred for the CoCr proximal component when compared to SS component.

It was surprising to find that the hard CoCr test component exhibited a weight loss, and particularly that it showed a greater weight loss than the much softer UHMWPE test component. From a tribological point of view this is an unexpected result as it is the softer component which is expected to wear. Moreover the roughness results showed that the CoCr articulating surfaces were relatively unchanged in terms of their Sa values during and at the end of five million cycles of testing. What could explain this inconsistency between gravimetric results indicating wear and roughness results indicating no wear? To try and find an answer, the surfaces of the hydroxyapatite covered stems were investigated, with the test stems compared with control stems. The control stems exhibited a uniform coating of hydroxyapatite, whilst the test stems did not. Instead the micrograph images of the test stems revealed distinct areas where the hydroxyapatite had been removed (Figures 7 and 8). Therefore, an explanation for the decrease in weight of the test components over five million cycles of testing is that this decrease was linked to removal of hydroxyapatite. That this decrease only affected the test components and not the control components indicates that hydroxyapatite removal did not occur due to the cleaning and weighing protocol, but was instead associated with testing itself. This information could be vital to other researchers who undertake *in vitro* wear tests of artificial joints with hydroxyapatite covered stems.

One limitation of this study is the small sample size. One MatOrtho PIPR prosthesis pair was tested *in vitro*, with a further pair used to account for lubricant uptake. However, this matches the quantity used for *in vitro* testing of: Single piece silicone prostheses [6]; and for SS-UHMWPE prostheses [32], both of which were evaluated using the simulator featured in this present study. Five million test cycles were undertaken, which conforms to the recommended standards for testing hip [18] and knee [19] prostheses *in vitro*. This paper offers the first ever reported *in vitro* wear tests of a metal-on-polymeric "proximal interphalangeal" joint in the scientific literature. This is important as clinical failures of finger prostheses continue to occur [34–38]. It has been argued previously that appropriate *in vitro* testing of finger prostheses can help to improve the performance of these implants [39] and this paper offers further evidence in support of this ideal.

5. Conclusions

A commercially available design of PIP prosthesis was tested to five million cycles of flexion-extension in a clinically validated simulator. Surface roughness and visual results indicated that minimal wear of the articulating surfaces occurred. However, weight loss of the test components was found. This has been explained by the loss of material from the hydroxyapatite coated stems, rather than loss of material from the articulating surfaces.

Acknowledgments: The authors would like to thank Anthony Unsworth of Durham University for allowing the finger simulator to be used at Newcastle University for this study and for his comments on a draft of the manuscript. The prostheses evaluated in this study were supplied free of charge by the Upper Limb Unit, Wrightington hospital, Wigan, UK.

Author Contributions: Andrew Naylor conducted the experimental work and wrote the initial manuscript. Ian A. Trail and Sumedh C. Talwalkar provided the prostheses and reviewed the study from a clinical perspective. Thomas J. Joyce was the principal investigator, reviewed the manuscript and was one of the original designers of the finger simulator used for this study.

Conflicts of Interest: The MatOrtho PIPR was designed in collaboration with one of the authors of this study, Ian A. Trail.

References

1. Lawrence, J.S.; Bremner, J.M.; Bier, F. Osteo-arthrosis. Prevalence in the population and relationship between symptoms and x-ray changes. *Ann. Rheum. Dis.* **1966**, *25*, 1–24. [PubMed]
2. Tsai, C.-L.; Liu, T.-K. Osteoarthritis in women: Its relationship to estrogen and current trends. *Life Sci.* **1992**, *50*, 1739–1744. [CrossRef]
3. Linscheid, R.L. Implant arthroplasty of the hand: Retrospective and prospective considerations. *J. Hand Surg. Am.* **2000**, *25*, 796–816. [PubMed]
4. Ashworth, C.R.; Hansraj, K.K.; Todd, A.O.; Dukhram, K.M.; Ebramzadeh, E.; Boucree, J.B.; Hansraj, M.S. Swanson proximal interphalangeal joint arthroplasty in patients with rheumatoid arthritis. *Clin. Orthop. Relat. Res.* **1997**, 34–37. [CrossRef]
5. Bales, J.G.; Wall, L.B.; Stern, P.J. Long-term results of swanson silicone arthroplasty for proximal interphalangeal joint osteoarthritis. *J. Hand Surg. Am.* **2014**, *39*, 455–461. [CrossRef] [PubMed]
6. Joyce, T.J.; Unsworth, A. The design of a finger wear simulator and preliminary results. *Proc. Inst. Mech. Eng. H.* **2000**, *214*, 519–526. [CrossRef] [PubMed]
7. Joyce, T.J.; Unsworth, A. A literature review of "failures" of the swanson finger prosthesis in the metacarpophalangeal joint. *Hand Surg.* **2002**, *7*, 139–146. [CrossRef] [PubMed]
8. Takigawa, S.; Meletiou, S.; Sauerbier, M.; Cooney, W.P. Long-term assessment of swanson implant arthroplasty in the proximal interphalangeal joint of the hand. *J. Hand Surg. Am.* **2004**, *29*, 785–795. [CrossRef] [PubMed]
9. Broadbent, M.; Birch, A.; Trail, I. A short-term review of the finsbury proximal interphalangeal joint replacement. In Proceedings of the British Society for Surgery of the Hand, Autumn Scientific Meeting, Nottingham, UK, 12 November 2009.
10. Watts, A.; Trail, I. Anatomical small joint replacement of the hand. Available online: http://www.boneandjoint.org.uk/content/focus/anatomical-small-joint-replacement-hand (accessed on 9 February 2015).
11. Daecke, W.; Kaszap, B.; Martini, A.K.; Hagena, F.W.; Rieck, B.; Jung, M. A prospective, randomized comparison of 3 types of proximal interphalangeal joint arthroplasty. *J. Hand Surg. Am.* **2012**, *37*, 1770–1779. [CrossRef] [PubMed]
12. Johnstone, B.R. Proximal interphalangeal joint surface replacement arthroplasty. *Hand Surg.* **2001**, *6*, 1–11. [CrossRef] [PubMed]
13. Johnstone, B.R.; Fitzgerald, M.; Smith, K.R.; Currie, L.J. Cemented *versus* uncemented surface replacement arthroplasty of the proximal interphalangeal joint with a mean 5-year follow-up. *J. Hand Surg. Am.* **2008**, *33*, 726–732. [CrossRef] [PubMed]
14. Murray, P.M.; Linscheid, R.L.; Cooney, W.P.; Baker, V.; Heckman, M.G. Long-term outcomes of proximal interphalangeal joint surface replacement arthroplasty. *J. Bone Jt. Surg. Am.* **2012**, *94*, 1120–1128. [CrossRef]
15. Schindele, S.F.; Hensler, S.; Audigé, L.; Marks, M.; Herren, D.B. A modular surface gliding implant (capflex-pip) for proximal interphalangeal joint osteoarthritis: A prospective case series. *J. Hand Surg. Am.* **2015**, *40*, 334–340. [CrossRef] [PubMed]
16. Selier, J.G. *American Society for Surgery of the Hand: Essentials of Hand Surger*; Lippincott Williams & Wilkins: Philadelphia, PA, USA, 2002.
17. Joyce, T.J.; Unsworth, A. A test procedure for artificial finger joints. *Proc. Inst. Mech. Eng. H.* **2002**, *216*, 105–110. [CrossRef] [PubMed]
18. British Standards Institute. *BS ISO 14242-3: Implants for Surgery. Wear of Total Hip-Joint Prostheses*; British Standards Institute: London, UK, 2009.
19. British Standards Institute. *BS ISO 14243-3: Implants for Surgery. Wear of Total Knee-Joint Prostheses*; British Standards Institute: London, UK, 2014.

20. Joyce, T.J. Prediction of lubrication regimes in two-piece metacarpophalangeal prostheses. *Med. Eng. Phys.* **2007**, *29*, 87–92. [CrossRef] [PubMed]
21. Joyce, T.J.; Rieker, C.; Unsworth, A. Comparative *in vitro* wear testing of peek and uhmwpe capped metacarpophalangeal prostheses. *Biomed. Mater. Eng.* **2006**, *16*, 1–10. [PubMed]
22. Naylor, A.; Bone, M.C.; Unsworth, A.; Talwalkar, S.C.; Trail, I.A.; Joyce, T.J. Evaluating the wear and surface roughness of pyrocarbon finger prostheses tested *in vitro*. In Proceedings of the 10th Anniversary Bath Biomechanics Symposium: Current Issues and Future Opportunities in Orthopaedic Research, Bath, UK, 15 September 2014.
23. Naylor, A.; Bone, M.C.; Unsworth, A.; Trail, I.A.; Talwalkar, S.C.; Joyce, T.J. *In vitro* testing of pyrolytic carbon proximal interphalangeal prostheses. In Proceedings of the British Society for Surgery of the Hand, Spring Scientific Meeting, Gateshead, UK, 1–2 May 2014.
24. Joyce, T.J.; Unsworth, A. Neuflex metacarpophalangeal prostheses tested *in vitro*. *Proc. Inst. Mech. Eng. H.* **2005**, *219*, 105–110. [CrossRef] [PubMed]
25. Joyce, T.J.; Unsworth, A. The wear of artificial finger joints using different lubricants in a new finger wear simulator. *Wear* **2001**, *250*, 199–205. [CrossRef]
26. Rempel, D.; Dennerlein, J.; Mote, C.D., Jr.; Armstrong, T. A method of measuring fingertip loading during keyboard use. *J. Biomech.* **1994**, *27*, 1101–1104. [CrossRef] [PubMed]
27. Jones, A.R.; Unsworth, A.; Haslock, I. A microcomputer controlled hand assessment system used for clinical measurement. *Eng. Med.* **1985**, *14*, 191–198. [CrossRef] [PubMed]
28. Weightman, B.; Amis, A.A. Finger joint force predictions related to design of joint replacements. *J. Biomed. Eng.* **1982**, *4*, 197–205. [CrossRef] [PubMed]
29. Katti, K.S. Biomaterials in total joint replacement. *Colloids Surf. B Biointerfaces* **2004**, *39*, 133–142. [CrossRef] [PubMed]
30. Szycher, M. *Biocompatible Polymers, Metals, and Composites*; Technomic Pub. Co.: Lancaster, OA, USA, 1983.
31. Joyce, T.J.; Langton, D.J.; Jameson, S.S.; Nargol, A.V.F. Tribological analysis of failed resurfacing hip prostheses and comparison with clinical data. *Proc. Inst. Mech. Eng. J.* **2009**, *223*, 317–323. [CrossRef]
32. Joyce, T.J. Wear testing of a djoa finger prosthesis *in vitro*. *J. Mater. Sci. Mater. Med.* **2010**, *21*, 2337–2343. [CrossRef] [PubMed]
33. Joyce, T.J.; Unsworth, A. The influence of bovine serum lubricant on the wear of cross-linked polyethylene finger prostheses. *J. Appl. Biomater. Biomech. JABB* **2004**, *2*, 136–142.
34. Bone, M.C.; Cunningham, J.L.; Lord, J.; Giddins, G.; Field, J.; Joyce, T.J. Analysis of failed van straten lpm proximal interphalangeal prostheses. *J. Hand Surg. Eur. Vol.* **2013**, *38*, 313–320. [CrossRef] [PubMed]
35. Hernández-Cortés, P.; Pajares-López, M.; Robles-Molina, M.J.; Gómez-Sánchez, R.; Toledo-Romero, M.A.; de Torres-Urrea, J. Two-year outcomes of elektra prosthesis for trapeziometacarpal osteoarthritis: A longitudinal cohort study. *J. Hand Surg. Eur. Vol.* **2012**, *37*, 130–137. [CrossRef] [PubMed]
36. Hobby, J.L.; Edwards, S.; Field, J.; Giddins, G. A report on the early failure of the LPM proximal interphalangeal joint replacement. *J. Hand Surg. Eur. Vol.* **2008**, *33*, 526–527. [CrossRef] [PubMed]
37. Kaszap, B.; Daecke, W.; Jung, M. High frequency failure of the moje thumb carpometacarpal joint arthroplasty. *J. Hand Surg. Eur. Vol.* **2012**, *37*, 610–616. [CrossRef] [PubMed]
38. Middleton, A.; Lakshmipathy, R.; Irwin, L.R. Failures of the rm finger prosthesis joint replacement system. *J. Hand Surg. Eur. Vol.* **2011**, *36*, 599–604. [CrossRef] [PubMed]
39. Joyce, T.J. Snapping the fingers. *J. Hand Surg. Br.* **2003**, *28*, 566–567. [CrossRef] [PubMed]

 *lubricants*

Review

Materials and Their Failure Mechanisms in Total Disc Replacement

John Reeks and Hong Liang *

Materials Science and Engineering, Texas A&M University, College Station, TX 77843-3123, USA; jreeks@tamu.edu
* Author to whom correspondence should be addressed; hliang@tamu.edu; Tel.: +1-979-862-2623; Fax: +1-979-845-0381.

Academic Editors: J. Philippe Kretzer and Amir Kamali
Received: 13 February 2015; Accepted: 14 April 2015; Published: 28 April 2015

Abstract: Adults suffering from lower back pain often find the cause of pain is degenerative disc disease. While non-surgical treatment is preferred, spinal fusion and total disc replacement remain surgical options for the patient. Total disc replacement is an emerging and improving treatment for degenerative discs. This paper provides a review of lumbar disc replacement for treatment of lower back pain. The mechanics and configuration of the natural disc are first discussed, followed by an introduction of treatment methods that attempt to mimic these mechanics. Total disc replacement types, materials, and failure mechanisms are discussed. Failure mechanisms primarily involve biochemical reactions to implant wear, as well as mechanical incompatibility of the device with natural spine motion. Failure mechanisms include: osteolysis, plastic deformation of polymer components, pitting, fretting, and adjacent level facet and disc degeneration.

Keywords: total disc replacement; failure; implants; degenerative disc disease; osteolysis; biomaterials; wear

---

1. Introduction

Lower back pain (LBP) is among the most common ailments for adults in the United States. According to a national survey performed by the Center for Disease Control and Prevention (CDC), 28.8% of adults have complaints of lower back pain [1]. Many cases of LBP are caused by intervertebral disc (IVD) injury or degenerative disc disease (DDD), or other disc injury. The IVD relies on natural flexibility of its materials, as well as its pseudo-pneumatic structure to transmit loads down the spine even in awkward loading positions [2]. In addition to transmitting loads and maintaining structure, the natural IVD behaves as a "hydrodynamic ball bearing" [2] allowing for smooth rotational motion. Natural IVD permits angular movement in the three planes of motion (sagittal, lateral, and transverse) giving it six degrees of freedom (DOF) [2–4]. It is in this capacity that the IVD is unlike any other joint. The balance of motion and motion resistance allow the body to maneuver in a complex manner while minimizing the stress induced by other parts of the body. Unfortunately, its duty to simultaneously undergo torsion, shear stress, localized compressive stress, and localized tensile stress both causes DDD and creates problems for treatment thereof.

A simple model of the IVD is shown in Figure 1. It is composed of a nucleus pulposus (inner structure) and the annulus fibrosus (outer structure). The nucleus can absorb some impact between adjacent vertebral bodies, thus protecting the rigid vertebral bodies from each other. The pulposus also acts to preserve the disc itself by helping withstand compression [4]. Surrounding the nucleus is the lamellar [2,4] annulus fibrosus. The annulus is composed of layers of oriented collagen fibers [4,5]. The collagen fibers in the annulus are arranged such that every other layer's fibers have the same orientation. The adjacent layers, however have the opposite orientation [4,6–8]. These fibers are generally positioned at a 65° angle from the y axis in Figure 1 [4].

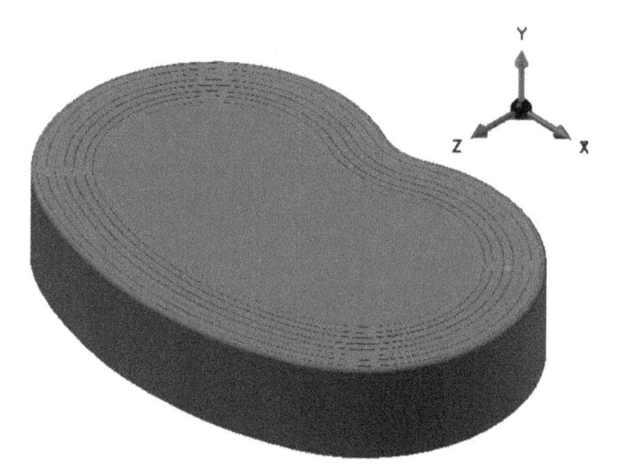

Figure 1. Drawing of the structure of an intervertebral disk.

One treatment for DDD is spinal fusion. This treatment method alleviates pain, but limits range of motion (ROM) and mobility for the patient. Total disc replacement (TDR) is another treatment method for DDD that aims to preserve motion and limit complications related to spinal fusion, such as adjacent level wear and disc degeneration [8–12]. This motion preservation is critical to preventing wear and injury in the area surrounding the treated disc. Through the inhibition of motion fusion creates stress concentrations at adjacent level facets and discs [9,12–14]. TDR is an effort to solve these issues and help maintain normal range or motion for patients [11].

Although the preservation of motion through TDR alleviates some problems associated with spinal fusion, TDR introduces different types of problems and failures. TDR issues and problems not seen in other implants vary based on the design of the TDR implant. These issues are primarily derived from the following: hyper-mobility, hypo-mobility, material wear, and wear debris particles. This paper will review failure mechanisms of TDR implants.

2. Lumbar Total Disc Replacements

To be effective, TDR implants must fulfill three requirements. The implant must have a solid, nondestructive interface with the adjacent vertebral bodies, provide mobility, and resist wear. More is required of lumbar than cervical TDR due to the extra loads it must bear. The lumbar spine supports more weight and encounters moments of greater magnitude than the cervical spine.

2.1. Types of Lumbar TDR

Lumbar total disc replacement can be classified according to their configuration as well as materials. The configurations of these devices are dependent upon the type of modules involved in the working disc. To allow maximum range of motion and permit the most freedom, current designs are built around a bearing. Bearings are one-piece (1P), Metal-on-Metal (MoM), or Metal-on-Polymer (MoP). MoM and MoP bearings use a ball and socket design to permit motion in all directions. Table 1 shows the different materials, classifications, and bearing designs for current TDRs.

Table 1. Summary of current total disc replacement (TDR) classification, materials, bearing type and regulatory status.

Device	Classification	Biomaterials	Bearing Design	References	Examples of Manufacturer
CHARITE	MoP	CoCr-UHMWPE	Mobile	[10,15–18]	DePuy Spine
Prodisc-L	MoP	CoCr-UHMWPE	Fixed	[15,19]	DePuy Synthes
Activ-L	MoP	CoCr-UHMWPE	Mobile	[20]	Aesculap
Mobidisc	MoP	CoCr-UHMWPE	Mobile	[10,21]	LDR Medical
Baguera	MoP	DLC coated Ti-UHMWPE	Fixed	[15]	Spineart
NuBac	PoP	PEEK-PEEK	Fixed	[22]	Pioneer
Maverick	MoM	CoCr-CoCr	Fixed	[15]	Medtronic
Kineflex	MoM	CoCr-CoCr	Mobile	[10,15]	SpinalMotion
Flexicore	MoM	CoCr-CoCr	Constrained		Stryker
XL-TDR	MoM	CoCr-CoCr	Fixed	[10,23]	NuVasive
CAdisc-L	One piece (1P)	PU-PC graduated modulus	1P	[10,15,24]	Rainier Technology
Freedom	1P	Ti plates; silicone PU-PC core	1P	[10,15]	Axiomed
eDisc	1P	Ti plates; elastomer core	1P	[10,15]	Theken
Physio-L	1P	Ti plates; elastomer core	1P	[10,15,25,26]	NexGen Spine
M6-L	1P	Ti plates; PU-PC core with UHMWPE fiber encapsulation	1P	[15]	Spinal Kinetics
LP-ESP (elastic spine pad)	1P	Ti endplates; PUPC coated silicone gel with microvoids	1P	[3]	FH Orthopedics

Descriptions in Table 1 also indicate bearing design as mobile, fixed, constrained, or 1P. Except for 1P bearings, the designs involve a ball and socket. A fixed bearing involves no moving parts except the sliding of the socket over the ball, but mobile and constrained bearings permit motion of the ball component.

2.1.1. Ball-and-Socket

When designing the ball and socket TDR, a major source of inspiration is the total hip replacement. Hip replacements use a ball and socket design but also experience higher stress and load concentrations than what is needed for TDR. Material selection, therefore, is often inspired by materials used in hip replacement prostheses. The sliding surfaces for the ball and socket bearings are composed of CoCr-CoCr for MoM designs and CoCr-UHMWPE for MoP designs. CoCr alloys are used because they have been found to produce less wear debris in knee and hip replacements [10]. Figure 2 is a drawing of the ball and socket mechanism in these devices to roughly illustrate how they work.

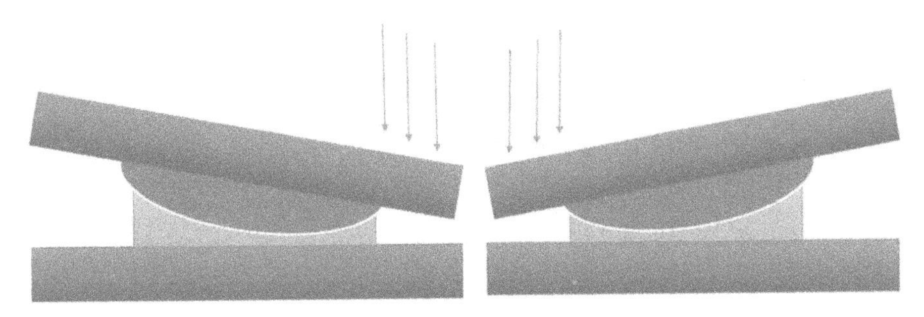

Figure 2. Illustration of working mechanisms of ball-socket devices.

2.1.2. One Piece

The one piece bearing design is a recent development for TDR. The aim of this design is to more adequately mimic the natural disc behavior through the implant. One piece designs reduce the number of surfaces on which wear can occur and they reduce hypermobility of the joint. Additionally, since these are made of a softer elastomer, they also mimic the cushioning provided by the natural intervertebral disc.

2.2. *Implant Materials*

Biomaterials being used in current implant designs are: cobalt-chromium alloy (CoCr), titanium (Ti), diamond-like carbon (DLC) coatings, ultrahigh-molecular-weight-polyethylene (UHMWPE), silicone, and a polyurethane-polycarbonate elastomer (PU-PC) [15,25]. Ti is typically used as the material to interface between implant and bone; the other materials listed are used as bearing materials for each design. Material selection is not entirely random; most of the early design ideas for TDR implants are a result of current designs in other total joint replacements. Bearing material selections are typically inspired by hip or knee replacement devices.

3. Causes for Failure

3.1. *Device Degradation*

Orthopedic implant failure can often be traced to the degradation of an implant. There are two mechanisms for the degradation of TDR implants: wear and corrosion [27]. This is to be expected with articulating bearings in harsh environments.

3.1.1. Metal-on-Polymer

When comparing mobile and fixed MoP devices, one can compare Prodisc-L with CHARITE TDR devices [28]. The mobile design in the CHARITE allows the UHMWPE core to shift as the back bends, which reduces stress experienced by posterior facets [28]. While this implies that the mobile bearing is a better design due to smaller loadings on facets and adjacent bodies, four degree of freedom (DOF) and 5DOF tests indicate higher UHMWPE wear mass loss in the CHARITE design [29,30].

Removed Mobidisc implants revealed that large UHMWPE particles surrounding the implant that caused an inflammatory response [21]. Particles were also found in patients with the Activ-L implant, but they were smaller and still triggered an inflammatory response [21]. Wear particles from UHMWPE tend to be spherical and ranging between 0.5 and 10 microns in diameter [31,32]. This means that even though UHMWPE is not chemically reactive and spherical particles are typically nonreactive as well, the large particle size would induce inflammatory and osteolytic responses [31–33].

In studies that have retrieved UHMWPE cores from used implants show measurable wear on the surface of the bearing [18,34]. As in hip replacements, multidirectional scratches and some penetration into the polymer were observed. Surface wear and damage are observed near the center of the core as a result of adhesive/abrasive mechanisms [18,34,35]. Some long term retrieval studies of the Prodisc-L also found third body wear and end plate impingement, often coupled with burnishing of the metallic endplates [36,37]. Impingement was also demonstrated in both Activ-L, CHARITE, and Mobidisc implant designs [21,38]. Retrieval studies also observed wear and damage that was similar to that of hip and knee implants. Central dome regions of the UHMWPE core exhibited microscopic scratches resembling the wear patterns of knee and hip MoP devices [35,36,39]. Scratch penetration into the polymer core increases at a decreasing rate with implantation time [34,40]. This behavior is similar to wear seen in hip implants. These studies, however, found this damage to be insignificant. The rim of the core was subject to extensive plastic deformation and fracture, which is sometimes observed in total knee replacements [34,41].

3.1.2. Metal-on-Metal

In hip arthroplasty it was noticed that MoP bearing surfaces produced high volume wear rates when compared to MoM devices. UHMWPE joints can wear at a rate over 100 microns per year [33]. MoM bearings were introduced to hip replacements to produce less wear debris and reduce particle-driven osteolysis [15,42,43]. However, in spite of the decrease in wear volume, the number of wear particles actually increased [44]. This is due to particle size differences in the materials. While MoP devices tend to wear at faster rates yielding more volume loss, MoM devices produce a large number of particles which range in size from 10 to 119 nm [45]. When compared to the particle size of UHMWPE particles on the order of microns in diameter, this means a substantial increase in the amount of reactive particles and reactive surface area of debris. This was especially damaging when the bearing was made of a reactive cobalt-based alloy, which is a major concern for the application of MoM bearings in TDR [15,44,46].

Although limiting volumetric wear is desirable, that is not all that must be considered. MoM bearings tend to produce more particles, which are much smaller than those produced by MoP bearings [44,46]. CoCr is used for these bearings because it has good mechanical properties [47,48] hard and passive in the biological environment [47,49–51]. It is not, however, inert. CoCr wear particles are chemically reactive within the body. This, combined with the large surface area of the debris, make the particles susceptible to electrochemical processes, which can lead to corrosion of the material and implant within the body [50,52,53]. Unlike the UHMWPE debris mentioned above, MoM bearing wear to produce fine, needle and fiber-shaped particles. This particle shape is chemically more reactive in nature, thus contributing to corrosion, tribocorrosion, and toxic and biological responses, such as metallosis, pseudotumors, biological reactions, osteolysis, and inflammation [15,44,46,54,55].

MoM implants reduce corrosion and negative biological response through passive oxide films [56,57]. Cyclic loading of these implants can lead to fretting which can disrupt oxide films and limit their effectiveness [52,53,58–60]. Corrosion in these implants is often a result of this fretting wear, which deteriorates the passive oxide layer. Corrosive wear usually produces particles of cobalt or chromium oxides or metal ions, which can build up or bind to proteins within the body causing severely negative biological response [27,42,43,52,60]. This response includes toxic reactions as well as metal build up and bone decay which are absolutely detrimental to patient health.

3.1.3. Polymer-on-Polymer

In addition to TDR, some Nubac, a PEEK-on-PEEK device proposes a nucleus replacement solution to minimize some of the biomechanical incompatibilities of TDR. This type of devices utilizes a PEEK-on-PEEK (PoP) bearing design, which replaces the nucleus pulposus. This design is not usually pursued because nucleus replacements have a tendency to be expelled by the body naturally. The Nubac device, however, showed a low risk of expulsion [22]. This device shows promise as, in addition to resisting expulsion, the fatigue resistance and wear resistance of PEEK. Preclinical cadaver and animal tests show that the device does not have detectible wear debris or any negative biological response to implantation [22]. There are issues, however, with device rejection and biological reactions. Long-term clinical trials showed an unacceptable rate of device migration with Nubac, which, in preclinical trials, showed little to no risk [61]. This migration was severe enough to have the device move into surrounding muscle tissue [61].

3.1.4. Diamond-Like Carbon

One disc design, the Baguera-L (See Table 1), moves to reduce the amount of metal wear debris by utilizing a DLC coating on the Ti endplate. DLC is known for its wear resistance and having low friction [62]. These properties make it an ideal candidate for ball and socket bearings in TDR. In spite of its superior wear resistance DLC has issues with long-term adhesion to the substrate.

DLC films, however, are prone to failure and delamination. Delamination primarily occurs due to corrosion of the substrate by body fluids that pass through pinholes in the DLC film. Delamination can also be driven by pitting corrosion under pinholes over time. A corrosive electrolyte may form and begin to corrode the substrate to fuel local debonding [62,63]. This local debonding can induce delamination through corrosion cracking (CC) and stress corrosion cracking (SCC) [62,64].

3.1.5. One Piece

An early design for a singular piece bearing in a TDR implant is the Acroflex disc design by Acromed Corporation. This design uses a polyolefin rubber to mimic the mechanical behavior of a natural disc. Although the device was tested for biological and biomechanical compatibility prior to clinical trials [25,65–68], patients suffered core material tears and failure mechanisms associated with fatigue [25]. Recent designs have employed PUPC as a core material which has a longer fatigue life.

Even with the improved elastomeric core, these devices could still fail due to PUPC tears or loss of adhesion between the different materials (PUPC-Ti, Ti-bone, or PUPC-bone). Since these designs are still fairly young, evaluating wear and corrosion resistance, as well as effectiveness, for these next-generation PUPC TDR designs require more long-term clinical testing [3,15,25].

3.2. The Body's Response

3.2.1. Biological Response

In TDR implants the biological response to wear debris is based on number of particles, particle size, particle shape, and chemical composition of the debris [27]. This indicates that biological responses are TDR-design and material dependent. Material-independent responses include inflammation and osteolysis. Osteolysis is often a result of the inflammatory response.

Osteolysis is a mode of degradation which involves the destruction of bone [69]. In the case of orthopedic implants and arthroplasty, osteolysis occurs at the interface between bone and implant. Osteolysis in TDR implants is primarily driven by micro-motion of the implant and the body's biological response to wear debris [33]. It is easy to picture how motion would cause bone degeneration: just like the periodic maneuvering of a shovel in soft clay will loosen and remove some clay. The osteolytic response to debris particles, however, is not so straight forward.

Bone growth and maintenance are the results of equilibrium responses of bone growth and resorption, which are driven by osteoblasts and osteoclasts, respectively. Debris particles disrupt bone homeostasis [27] through an inflammatory response which in turn stimulates the maturation of osteoclasts increasing bone resorption [27,33]. The combination of these wear processes leads to increased wear rates over time as resorption loosens the device, thus, creating more space for osteolysis-causing motion and debris [70].

Osteolysis is a prominent mode of failure for total joint replacements including TDR [71]. According to some studies osteolysis is not as prevalent TDR due to it having a smaller range of motion than hip and knee replacements, which would indicate smaller wear tracks and fewer debris particles. These sources reported that wear particles do remain a concern in spinal implants because they indirectly induce osteoclastogenesis which can lead to bone resorption and inflammation [15,59,72,73]. This limited wear of TDR, however, is not necessarily the case for MoM implants. A review of spine implant debris by Hallab reveals that though MoP hip implants have a substantially higher wear volumerate than TDR, MoM hip and intervertebral disc replacements have comparable wear rates [27].

Problematic responses to CoCr MoM wear particles include metal hypersensitivity, metallosis, the formation of pseudotumors, and vasculitis [15,44,46,74–76]. Metal-specific biological responses are a result of wear particle chemistry and shape. Co and Cr ions are chemically reactive and therefore are pro-inflammatory [27,45]. Aside from the chemical nature of CrCo debris, shape is an important factor dictating biological response. As seen historically with asbestos fibers, fiber shaped particles from TDR wear cause greater inflammatory response than round particles through a natural response called "danger signaling" [32,77–79].

3.2.2. Mechanical Response

A problem that separates TDR from other joint replacements lies in the complex duty of the intervertebral disc. It is not merely a joint which connects two moving parts. The intervertebral disc, especially in the lumbar region, serves to resist motion and reduce stress concentrations in the adjacent level tissue [4,9].

(a) Ball-and-Socket Bearing

Ball and socket bearings to not completely replicate the functions of the natural IVD. These bearings tend to be axially rigid and are not designed to resist bending or rotational moments, allowing motion to occur in excess of that permitted by the natural disc [9,80]. These qualities can lead to changes in range of motion (ROM), segmental lordosis, or facet stressing [3,15,26,81–83].

Among limited number of available reports, the Spine Institute at The Ohio State University (OSU) used a hybrid biomechanical model to analyze lumbar function after implantation of Prodisc-L TDR under different external loading conditions. This model was made after an individual male and uses a flexible multi-body dynamic analysis system [9]. The model looked at TDR at all lumbar levels, and examined vertebral body loading, ROM, and ligament and facet joint forces. These measurements were taken when the subject was not exerting at all and when he was asked to lift 9.5 and 19 kg separately. The figures below are drawn based on the OSU study. In the figures, the abbreviations SUP and INF represent the interface at the superior and inferior levels of the implant respectively. Figure 3 [9] shows mean values of data from a virtual model simulation of vertebral loadings from the OSU study at different implantation levels. The original data can be found in reference [9]. Before reviewing these data some terminology should be reviewed. Numbers such as L5 and S1 describe specific vertebrae. L5 is the fifth and most inferior lumbar (L) vertebrae. S1 is representative of the sacrum which is inferior to the spine and is immobile.

(b) Vertebral Body Stresses

Figure 3 below shows approximate mean value data for different simulated loadings based on disc type and implantation level. The information is organized such that data for a given implantation level are reported in the same color. The study found that the substantial differences in performance between TDR and the natural disc lied closer to L5/S1 level. Therefore, the information below focuses on that region. After examining Figure 3 (a) it can be noted that in general at these lower lumbar levels (L4/L5 to L5/S1) that compressive loads apply on the interfacing vertebral bodies. These loading changes are primarily due to high stiffness of the implant and discontinuity at the bone-implant interface. Similar trends of excess loadings due to TDR can be noted for vertebral body shear loadings as well. In the case of AP shear loading, seen in Figure 3 (c) the direction and magnitude of the shear are dependent upon the shape of the spine. Since the resistance to motion is less with TDR, less shear loading is exhibited with TDR than an intact disc in regions of spinal curvature. This, however would indicate higher loading of the nearby ligaments.

(c) ROM; Facet and Ligament Stresses

A study was conducted by Burger *et al.* to examine lumbar facet forces on cadavers after Prodisc-L TDR [84]. A facet is another joint at which adjacent vertebrae interface. The purpose of the facet joint is to aid in the inhibition of rotation and excess motion in order to keep vertebrae aligned. Each level has 2 facet joints, one on the left and one on the right side of the spinous process.

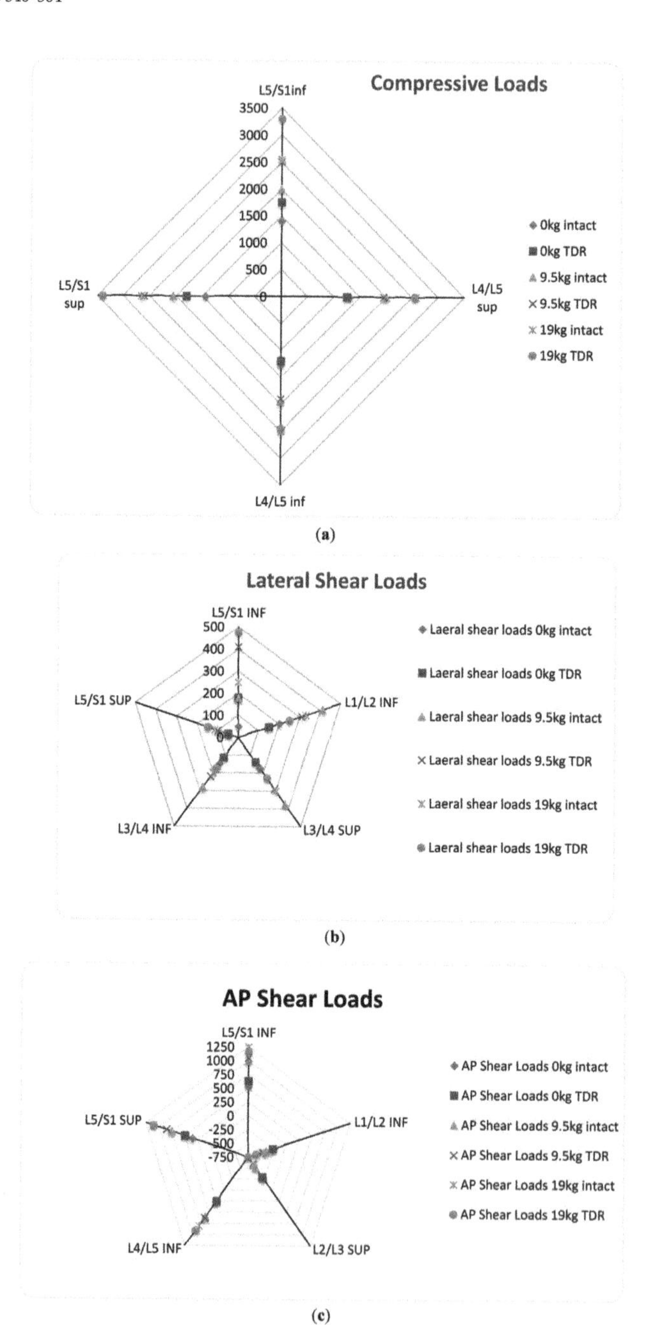

Figure 3. Selected mean data from a published simulated work in reference [9]. (**a**) Comparison of simulated compressive vertebral loads before and after TDR implantation, based on exterior loading and implantation level; (**b**) Comparison of simulated anterior-posterior shear loading at vertebral body interface before and after TDR implantation based exterior loading and implantation level; (**c**) Comparison of simulated lateral shear loading at vertebral body interface before and after TDR implantation based on exterior loading and implantation level [9].

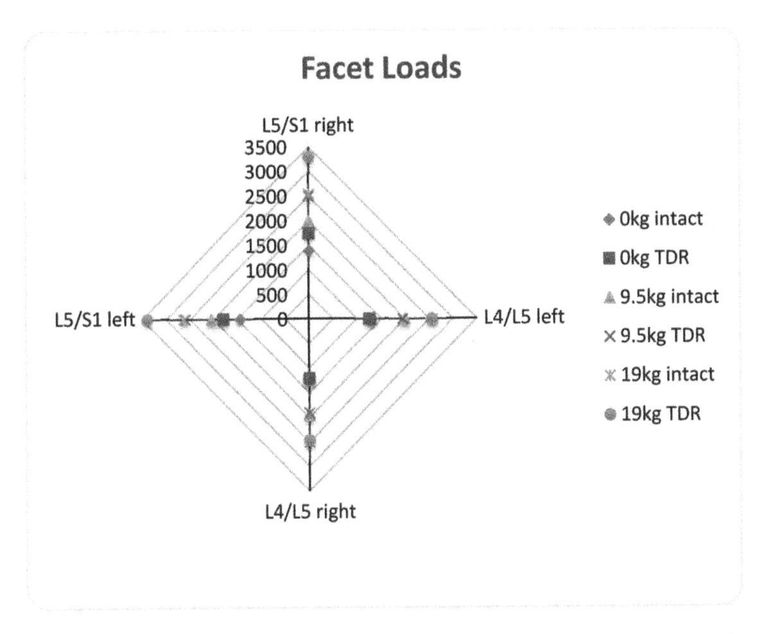

Figure 4. Selected data from simulated facet load comparison of natural disc to TDR implantation based on implantation level and exterior loading [9].

Figure 4 shows mean adjacent facet loads for TDR at each level according to the OSU study. Once again, facet loading most substantially increases with TDR implantation at L5/S1 level. Additionally, the OSU study shows increased ligament loading and ROM in all directions specifically at L4/L5 and L5/S1 levels of implantation [9]. The most drastic ROM difference occurred in the transverse plane while the motion was twisting about the spine's axis [9].

Cadaveric study performed has also examined lumbar implantation. Similarly, those concluded that lateral bending and axial rotation result in the most significant increases in facet loading [38,84,85].

This increase in facet loading and ROM can cause stress fractures of the pars and bilateral pedicle, as well as adjacent level disc degeneration [25,86–89]. This indicates much need for improvement of TDR implants in the mechanical regime. The articulating ball and socket joints, while they soothe pain, tend to be a source of mechanical issues and injury risk for patients.

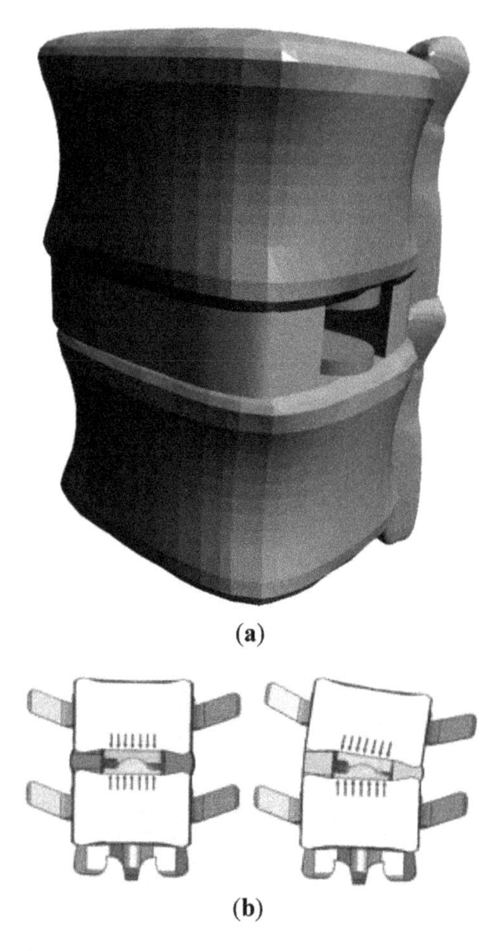

Figure 5. (**a**) Drawing of XL-TDR implanted at L4/L5 level shows retention of natural disc tissue and structure with lateral TDR implantation [90]; (**b**) Illustration of Nubac nucleus replacement (used with permission) [22].

Two possible solutions for the mechanical incompatibilities as shown above are: 1P bearings and partial replacements like Nubac and the extreme lateral TDR (XL-TDR) by NuVasive. The XL-TDR is designed to limit the amount of natural tissue removed. This allows the body to retain some of its natural cushioning and resistance to motion. Figure 5a shows how the lateral insertion can eliminate some need for tissue removal. Figure 5b is an illustration of the nucleus replacement and how it helps preserve natural motion.

The XL-TDR design improves on other MoM implants by preventing hypermobility, providing some natural biomechanical support, and provides a unique look at modern TDR implant designs [90].

4. One Piece

In order to correct problems seen with ball and socket TDR implants, the one piece bearing employs an elastomer to mimic the body's natural motion [3,15,83]. The current implant designs using the PUPC bearing passed fatigue and endurance bending tests prior to introduction into clinical trials [25]. These cores were found to resist tearing and wear and were found to be more biomechanically compatible with natural motions [25]. As mentioned before, these devices must be studied further in more long term trials to effectively understand the body's response.

Table 2 below gives a brief summary of failure mechanisms based on implant materials and design.

Table 2. Summary of common problems of different implant materials and their effects leading to device failure.

Bearing Type	Material	Problems	Effects	Reference
Ball and socket	CoCr	Reactive wear ions Fibrous particules	Metal sensitivity reactions, Inflammation, Osteolysis	[19,31,68,91,92]
		Metallosis		[15,44,61]
		No shock absorption	Compressive stresses on vertebral bodies	[9]
	UHMWPE	Wear debris large wear volume	Bone resorption, Osteolysis	[31,92,93]
				[18,31]
		Plastic deformation		[18]
		Increased ROM	Facet and ligament loading	[16]
	PEEK	No shock absorption	Compressive stresses on vertebral bodies	[9]
		Prosthesis migration	Biomechanical incompatibility, stress on remaining annulus, total ejection of device	[61]
		Endplate reaction	Severe biological rejection	[61]
1P	PUPC	More studies necessary		[25]

5. Future Prospects

TDR looks to be a suitable treatment for lumbar DDD. More specifically, studies show that the elastomeric 1P bearing TDR are the future of DDD treatment [94,95]. These 1P bearings provide a solution to common problems seen with ball and socket TDR implants [20]. A hurdle for these designs is finding suitable core materials for biocompatibility and adhesion to the vertebral bodies [25]. This problem seems to have been solved by the use of the PUPC cores in the one piece (1P) bearing implant design, but more research still needs to be done for these designs [15]. Although the XL-TDR provides a solution to some of the problems with the traditional ball and socket design, they still involve articulating metal surfaces which can be the site of wear and corrosion. For this reason, research indicates that the 1P bearing would be the ideal design for the future of TDR. Perhaps even a 1P design placed laterally could further improve the TDR implant design.

6. Conclusions

Total disc replacement (TDR) are an effective solution to degenerative disc disease (DDD), but currents designs still require improvement to be better substitutes for a healthy disc. Ball and socket bearing designs, or any implant involving articulating surfaces increase the risk of failure due to an introduction of more wear and corrosion surfaces. These articulating bearings also do not adequately resist motion and provide cushioning or stability for the lumbar spine (with the exception of the XL-TDR). This would imply that the future of TDR employs a flexible elastomer bearing which can better mimic a natural intervertebral disc (IVD). While this seems like a natural solution, this design type must be looked into further and studied more closely before any conclusions can be drawn.

Acknowledgments: Authors wish to thank Alex Fang for support during preparation of this manuscript. We also wish to acknowledge Dongsheng Zhou and Guodong Wang of Shandong University Orthopedic Hospital for providing surgical images.

Author Contributions: John Reeks conducted review, analyzed data, and drafted the manuscript. Hong Liang directed review and wrote the paper.

Conflicts of Interest: The author declares that there is no conflict of interest.

References

1. Pleis, J.; Schiller, J.; Benson, V. Summary health statistics for U.S. adults: National Health Interview Survey, 2010. *Vital Health Stat.* **2012**, *10*, 1–207.
2. Karajan, N. Multiphasic intervertebral disc. mechanics: Theory and application. *Arch. Comput. Methods Eng.* **2012**, *19*, 261–339. [CrossRef]
3. Lazennec, J.Y.; Even, J.; Skalli, W.; Rakover, J.P.; Brusson, A.; Rousseau, M.A. Clinical outcomes, radiologic kinematics, and effects on sagittal balance of the 6 df LP-ESP lumbar disc prosthesis. *Spine J.* **2014**, *14*, 1914–1920. [CrossRef] [PubMed]
4. Neumann, D.A. *Kinesiology of the Musculoskeletal System: Foundations for Rehabilitation*; Elsevier Health Sciences: St. Louis, MO, USA, 2013.
5. Bogduk, N. *Clinical Anatomy of the Lumbar Spine and Sacrum*; Churchill Livingstone: London, UK, 2005.
6. Markolf, K.L.; Morris, J.M. The structural components of the intervertebral disc. *J. Bone Joint Surg.* **1974**, *56*, 675–687. [PubMed]
7. Rothman, R.H.; Simeone, F.A. The spine. *J. Pediatr. Orthop.* **1992**, *12*, 549. [CrossRef]
8. Gloria, A.; De Santis, R.; Ambrosio, L.; Causa, F.; Tanner, K.E. A multi-component fiber-reinforced PHEMA-based hydrogel/HAPEX device for customized intervertebral disc prosthesis. *J. Biomater. Appl.* **2011**, *25*, 795–810. [CrossRef] [PubMed]
9. Knapik, G. Use of a personalized hybrid biomechanical model to assess change in lumbar spine function with a TDR compared to an intact spine. *Eur. Spine J.* **2012**, *21*, 641–652. [CrossRef]
10. Serhan, H.; Mhatre, D.; Defossez, H.; Bono, C.M. Motion-preserving technologies for degenerative lumbar spine: The past, present, and future horizons. *SAS J.* **2011**, *5*, 75–89. [CrossRef] [PubMed]
11. Gornet, M.F.; Burkus, J.K.; Dryer, R.F.; Peloza, J.H. Lumbar disc arthroplasty with Maverick disc *versus* stand-alone interbody fusion: A prospective, randomized, controlled, multicenter investigational device exemption trial. *Spine* **2011**, *36*, E1600–E1611. [CrossRef] [PubMed]
12. Kumar, M. Correlation between sagittal plane changes and adjacent segment degeneration following lumbar spine fusion. *Eur. Spine J.* **2001**, *10*, 314–319. [CrossRef] [PubMed]
13. Kumar, M. Long-term follow-up of functional outcomes and radiographic changes at adjacent levels following lumbar spine fusion for degenerative disc disease. *Eur. Spine J.* **2001**, *10*, 309–313. [CrossRef] [PubMed]
14. Cakir, B.; Richter, M.; Schmoelz, W.; Schmidt, R.; Reichel, H.; Wilke, H.J. Resect or not to resect: The role of posterior longitudinal ligament in lumbar total disc replacement. *Eur. Spine J.* **2012**, *21*, 592–598. [CrossRef]
15. Veruva, S.Y.; Steinbeck, M.J.; Toth, J.; Alexander, D.D.; Kurtz, S.M. Which design and biomaterial factors affect clinical wear performance of total disc replacements? A systematic review. *Clin. Orthop. Related Res.* **2014**, *472*, 3759–3769. [CrossRef]
16. Takigawa, T.; Espinoza Orías, A.A.; An, H.S.; Gohgi, S.; Udayakumar, R.K.; Sugisaki, K.; Natarajan, R.N.; Wimmer, M.A.; Inoue, N. Spinal kinematics and facet load transmission after total disc replacement. *Spine* **2010**, *35*, E1160–E1166. [CrossRef] [PubMed]
17. Kettler, A.; Bushelow, M.; Wilke, H.J. Influence of the loading frequency on the wear rate of a polyethylene-on-metal lumbar intervertebral disc replacement. *Eur. Spine J.* **2012**, *21*, S709–S716. [CrossRef] [PubMed]
18. Shkolnikov, Y.P.; Bowden, A.; MacDonald, D.; Kurtz, S.M. Wear pattern observations from TDR retrievals using autoregistration of voxel data. *J. Biomed. Mater. Res. B Appl. Biomater.* **2010**, *94*, 312–317. [PubMed]
19. Chen, W.M.; Park, C.; Lee, K.; Lee, S. *In situ* contact analysis of the prosthesis componentsof Prodisc-L in lumbar spine following total disc replacement. *Spine* **2009**, *34*, E716–E723. [CrossRef] [PubMed]
20. Huang, R.C.; Girardi, F.P.; Cammisa, F.P.; Wright, T.M. The implications of constraint in lumbar total disc replacement. *J. Spinal Disord. Tech.* **2003**, *16*, 412–417. [CrossRef] [PubMed]
21. Austen, S.; Punt, I.M.; Cleutjens, J.P.; Willems, P.C.; Kurtz, S.M.; MacDonald, D.W.; van Rhijn, L.W.; van Ooij, A. Clinical, radiological, histological and retrieval findings of Activ-L and Mobidisc total disc replacements: A study of two patients. *Eur. Spine J.* **2012**, *21*, S513–S520. [CrossRef] [PubMed]

22. Bao, Q.B.; Songer, M.; Pimenta, L.; Werner, D.; Reyes-Sanchez, A.; Balsano, M.; Agrillo, U.; Coric, D.; Davenport, K.; Yuan, H. Nubac disc arthroplasty: Preclinical studies and preliminary safety and efficacy evaluations. *SAS J.* **2007**, *1*, 36–45. [CrossRef] [PubMed]
23. Marchi, L.; Oliveira, L.; Coutinho, E.; Pimenta, L. The importance of the anterior longitudinal ligament in lumbar disc arthroplasty: 36-Month follow-up experience in extreme lateral total disc replacement. *Int. J. Spine Surg.* **2012**, *6*, 18–23. [CrossRef] [PubMed]
24. McNally, D.; Naylor, J.; Johnson, S. An *in vitro* biomechanical comparison of Cadisc™-L with natural lumbar discs in axial compression and sagittal flexion. *Eur. Spine J.* **2012**, *21*, 612–617. [CrossRef]
25. Pimenta, L.; Springmuller, R.; Lee, C.K.; Oliveira, L.; Roth, S.E.; Ogilvie, W.F. Clinical performance of an elastomeric lumbar disc replacement: Minimum 12 months follow-up. *SAS J.* **2010**, *4*, 16–25. [CrossRef] [PubMed]
26. Van den, B. Design of next generation total disk replacements. *J. Biomech.* **2012**, *45*, 134–140. [CrossRef] [PubMed]
27. Hallab, N. A review of the biologic effects of spine implant debris: Fact. from fiction. *SAS J.* **2009**, *3*, 143–160. [CrossRef] [PubMed]
28. Frelinghuysen, P.; Huang, R.C.; Girardi, F.P.; Cammisa, F.P. Lumbar total disc replacement part I: Rationale, biomechanics, and implant types. *Orthop. Clin. North Am.* **2005**, *36*, 293–299. [CrossRef] [PubMed]
29. Vicars, R.; Hyde, P.J.; Brown, T.D.; Tipper, J.L.; Ingham, E.; Fisher, J.; Hall, R.M. The effect of anterior-posterior shear load on the wear of ProDisc-L TDR. *Eur. Spine J.* **2010**, *19*, 1356–1362. [CrossRef] [PubMed]
30. Vicars, R.; Prokopovich, P.; Brown, T.D.; Tipper, J.L.; Ingham, E.; Fisher, J.; Hall, R.M. The effect of anterior-posterior shear on the wear of CHARITE total disc replacement. *Spine* **2012**, *37*, E528–E534. [CrossRef] [PubMed]
31. Taki, N.; Tatro, J.M.; Nalepka, J.L.; Togawa, D.; Goldberg, V.M.; Rimnac, C.M.; Greenfield, E.M. Polyethylene and titanium particles induce osteolysis by similar, lymphocyte-independent, mechanisms. *J. Orthop. Res.* **2005**, *23*, 376–383. [CrossRef] [PubMed]
32. Sieving, A. Morphological characteristics of total joint arthroplasty-derived ultra-high molecular weight polyethylene (UHMWPE) wear debris that provoke inflammation in a murine model of inflammation. *J. Biomed. Mater. Res.* **2003**, *64*, 457–464. [CrossRef]
33. Agarwal, S. Osteolysis—basic science incidence and diagnosis. *Curr. Orthop.* **2004**, *18*, 220–231. [CrossRef]
34. Kurtz, S.M.; van Ooij, A.; Ross, R.; de Waal Malefijt, J.; Peloza, J.; Ciccarelli, L.; Villarraga, M.L. Polyethylene wear and rim fracture in total disc arthroplasty. *Spine J.* **2007**, *7*, 12–21. [CrossRef] [PubMed]
35. McKellop, H.A.; Campbell, P.; Park, S.H.; Schmalzried, T.P.; Grigoris, P.; Amstutz, H.C.; Sarmiento, A. The origin of submicron polyethylene wear debris in total hip arthroplasty. *Clin. Orthop. Related Res.* **1995**, *311*, 3–20.
36. Lebl, D. *In vivo* functional performance of failed Prodisc-L devices: Retrieval analysis of lumbar total disc replacements. *Spine* **2012**, *37*, E1209–E1217. [CrossRef] [PubMed]
37. Choma, T. Retrieval analysis of a ProDisc-L total disc replacement. *J. Spinal Disord. Tech.* **2009**, *22*, 290–296. [CrossRef] [PubMed]
38. Rundell, S.A.; Day, J.S.; Isaza, J.; Guillory, S.; Kurtz, S.M. Lumbar total disc replacement impingement sensitivity to disc height distraction, spinal sagittal orientation, implant position, and implant lordosis. *Spine* **2012**, *37*, E590–E598. [CrossRef] [PubMed]
39. Wang, A.; Essner, A.; Polineni, V.K.; Stark, C.; Dumbleton, J.H. Lubrication and wear of ultra-high molecular weight polyethylene in total joint replacements. *Tribol. Int.* **1998**, *31*, 17–33. [CrossRef]
40. Charnley, J.; Halley, D.K. Rate of wear in total hip replacement. *Clin. Orthop. Related Res.* **1975**, *112*, 170–179.
41. Wright, T.; Bartel, D. The problem of surface damage in polyethylene total knee components. *Clin. Orthop. Related Res.* **1986**, *205*, 67–74.
42. Gornet, M. Prospective study on serum metal levels in patients with metal-on-metal lumbar disc arthroplasty. *Eur. Spine J.* **2013**, *22*, 741–746. [CrossRef] [PubMed]
43. Kurtz, S. The latest lessons learned from retrieval analyses of ultra-high molecular weight polyethylene, metal-on-metal, and alternative bearing total disc replacements. *Semin. Spine Surg.* **2012**, *24*, 57–70. [CrossRef] [PubMed]
44. Guyer, R.D.; Shellock, J.; MacLennan, B.; Hanscom, D.; Knight, R.Q.; McCombe, P.; Jacobs, J.J.; Urban, R.M.; Bradford, D.; Ohnmeiss, D.D. Early failure of metal-on-metal artificial disc prostheses associated with

lymphocytic reaction: Diagnosis and treatment experience in four cases. *Spine* **2011**, *36*, E492–E497. [CrossRef] [PubMed]
45. Behl, B.; Papageorgiou, I.; Brown, C.; Hall, R.; Tipper, J.L.; Fisher, J.; Ingham, E. Biological effects of cobalt-chromium nanoparticles and ions on dural fibroblasts and dural epithelial cells. *Biomaterials* **2013**, *34*, 3547–3558. [CrossRef] [PubMed]
46. Golish, S.R.; Anderson, P.A. Bearing surfaces for total disc arthroplasty: Metal-on-metal *versus* metal-on-polyethylene and other biomaterials. *Spine J.* **2012**, *12*, 693–701. [CrossRef] [PubMed]
47. Gotman, I. Characteristics of metals used in implants. *J. Endourol.* **1997**, *11*, 383–389. [CrossRef] [PubMed]
48. Asphahani, A.I.; Kumar, P.; Hickl, A.J.; Lawley, A. Properties and characteristics of cast, wrought, and powder metallurgy (P/M) processed cobalt-chromium-molybdenum implant materials. In *Corrosion and Degradation of Implant Materials: Second Symposium*; Fraker, A.C., Griffin, C.D., Eds.; ASTM International: West Conshohocken, PA, USA, 1985.
49. Perkins, L. Evaluation of Bone Fixation Implants. Texas A&M University, College Station, TX, USA, December 2012.
50. Gurappa, I. Characterization of different materials for corrosion resistance under simulated body fluid conditions. *Mater. Charact.* **2002**, *49*, 73–79. [CrossRef]
51. Gurrappa, I. Corrosion and its importance in selection of materials for biomedical applications. *Corros. Prev. Control* **2001**, *48*, 23–37.
52. Taksali, S. Material considerations for intervertebral disc replacement implants. *Spine J.* **2004**, *4*, 231S. [CrossRef] [PubMed]
53. Pourbaix, M. Electrochemical corrosion of metallic biomaterials. *Biomaterials* **1984**, *5*, 122–134. [CrossRef] [PubMed]
54. Kirkpatrick, J. Corrosion on spinal implants. *J. Spinal Disord. Tech.* **2005**, *18*, 247–251. [PubMed]
55. Michel, R. Trace element burdening of human tissues due to the corrosion of hip-joint prostheses made of cobalt-chromium alloys. *Arch. Orthop. Trauma. Surg.* **1984**, *103*, 85–95. [CrossRef] [PubMed]
56. Pound, B.G. Passive films on metallic biomaterials under simulated physiological conditions. *J. Biomed. Mater. Res. Part A* **2014**, *102*, 1595–1604. [CrossRef]
57. Jacobs, J.J. Corrosion of metal orthopaedic implants. *JBJS J. Bone Joint Surg.* **1998**, *80*, 268–282.
58. Jacobs, J.J.; Shanbhag, A.; Glant, T.T.; Black, J.; Galante, J.O. Wear debris in total joint replacements. *J. Am. Acad. Orthop. Surg.* **1994**, *2*, 212–220. [PubMed]
59. Waterhouse, R.B. Fretting wear. *Wear* **1984**, *100*, 107–118. [CrossRef]
60. Zeh, A.; Becker, C.; Planert, M.; Lattke, P.; Wohlrab, D. Time-dependent release of cobalt and chromium ions into the serum following implantation of the metal-on-metal Maverick type artificial lumbar disc (Medtronic Sofamor Danek). *Arch. Orthop. Trauma Surg.* **2009**, *129*, 741–746. [CrossRef] [PubMed]
61. Pimenta, L. Lessons learned after 9 years' clinical experience with 3 different nucleus replacement devices. *Semin. Spine Surg.* **2012**, *24*, 43–47. [CrossRef]
62. Hauert, R.; Thorwarth, K.; Thorwarth, G. An overview on diamond-like carbon coatings in medical applications. *Surf. Coat. Technol.* **2013**, *233*, 119–130. [CrossRef]
63. Chandra, L. The effect of biological fluids on the adhesion of diamond-like carbon films to metallic substrates. *Diamond Related Mater.* **1995**, *4*, 852–856. [CrossRef]
64. Falub, C.V. *In vitro* studies of the adhesion of diamond-like carbon thin films on CoCrMo biomedical implant alloy. *Acta Mater.* **2011**, *59*, 4678–4689. [CrossRef]
65. Enker, P. Artificial disc replacement. Preliminary report with a 3-year minimum follow-up. *Spine* **1993**, *18*, 1061–1070. [CrossRef] [PubMed]
66. Serhan, H.; Ross, R.; Lowery, G.; Fraser, R. Biomechanical characterization of a new lumbar disc prosthesis. *J. Bone Joint Surg. Br.* **2002**, *84*, 215.
67. Moore, R. The biologic response to particles from a lumbar disc prosthesis. *Spine* **2002**, *27*, 2088–2094. [CrossRef] [PubMed]
68. Fraser, R. AcroFlex design and results. *Spine J.* **2004**, *4*, S245–S251. [CrossRef]
69. Credo Reference (Firm). *Merriam-Webster's Medical Desk Dictionary*; Thomson Delmar Learning: Clifton Park, NY, USA, 2006; p. 1.
70. Per Aspenberg, P.H. Periprosthetic bone resorption. Particles *versus* movement. *J. Bone Joint Surg. Br.* **1996**, *78*, 641–646. [PubMed]

71. Kang, J. Chronic failure of a lumbar total disc. Replacement with osteolysis. report of a case with nineteen-year follow-up. *JBJS J. Bone Joint Surg.* **2008**, *90*, 2230.
72. Bisseling, P.; Zeilstra, D.J.; Hol, A.M.; van Susante, J.L.C. Metal ion levels in patients with a lumbar metal-on-metal total disc replacement SHOULD WE BE CONCERNED? *J. Bone Joint Surg. Br.* **2011**, *93*, 949–954. [CrossRef] [PubMed]
73. Black, J. *In Vivo Corrosion of a Cobalt-Base Alloy and Its Biological Consequences, in Biocompatibility of Co-Cr-Ni Alloys*; Springer: New York, NY, USA, 1988; pp. 83–100.
74. Cavanaugh, D. Delayed hyper-reactivity to metal ions after cervical disc arthroplasty: A case report and literature review. *Spine* **2009**, *34*, E262–E265. [CrossRef] [PubMed]
75. Shang, X.; Wang, L.; Kou, D.; Jia, X.; Yang, X.; Zhang, M.; Tang, Y.; Wang, P.; Wang, S.; Xu, Y.; Wang, H. Metal. hypersensitivity in patient with posterior lumbar spine fusion: A case report and its literature review. *BMC Musculoskelet. Disord.* **2014**, *15*, 314. [CrossRef] [PubMed]
76. Jacobs, J.J.; Hallab, N.J.; Urban, R.M.; Wimmer, M.A. Wear particles. *J. Bone Joint Surg.* **2006**, *88*, 99–102. [CrossRef] [PubMed]
77. Laquerriere, P. Importance of hydroxyapatite particles characteristics on cytokines production by human monocytes *in vitro*. *Biomaterials* **2003**, *24*, 2739–2747. [CrossRef] [PubMed]
78. Bruch, J. Response of cell cultures to asbestos fibers. *Environ. Health Perspect.* **1974**, *9*, 253. [CrossRef] [PubMed]
79. Dostert, C. Innate immune activation through Nalp3 inflammasome sensing of asbestos and silica. *Science* **2008**, *320*, 674–677. [CrossRef] [PubMed]
80. Botolin, S. Facet joint biomechanics at the treated and adjacent levels after total disc replacement. *Spine* **2011**, *36*, E27–E32. [PubMed]
81. Chung, S. Biomechanical effect of constraint in lumbar total disc replacement: A study with finite element analysis. *Spine* **2009**, *34*, 1281–1286. [CrossRef] [PubMed]
82. O'Leary, P.; Nicolakis, M.; Lorenz, M.A.; Voronov, L.I.; Zindrick, M.R.; Ghanayem, A.; Havey, R.M.; Carandang, G.; Sartori, M.; Gaitanis, I.; *et al.* Response of CHARITE total disc replacement under physiologic loads: Prosthesis component motion patterns. *Spine J.* **2005**, *5*, 590–599. [CrossRef] [PubMed]
83. Heuer, F.; Schmidt, H.; Klezl, Z.; Claes, L.; Wilke, H.J. Stepwise reduction of functional spinal structures increase range of motion and change lordosis angle. *J. Biomech.* **2007**, *40*, 271–280. [CrossRef] [PubMed]
84. Botolin, S.; Puttlitz, C.; Baldini, T.; Petrella, A.; Burger, E.; Abjornson, C.; Patel, V. Facet joint biomechanics at the treated and adjacent levels after total disc replacement. *J. Bone Joint Surg. Br.* **2012**, *94*, 143–143.
85. Rundell, S. Total disc replacement positioning affects facet contact forces and vertebral body strains. *Spine* **2008**, *33*, 2510–2517. [CrossRef] [PubMed]
86. Phillips, F.; Diaz, R.; Pimenta, L. The fate of the facet joints after lumbar total disc replacement: A clinical and MRI study. *Spine J.* **2005**, *5*, S75. [CrossRef]
87. Shim, C.S. CHARITE *versus* ProDisc—A comparative study of a minimum 3-year follow-up. *Spine* **2007**, *32*, 1012–1018. [CrossRef] [PubMed]
88. Mathew, P. Bilateral pedicle fractures following anterior dislocation of the polyethylene inlay of a ProDisc artificial disc replacement: A case report of an unusual complication. *Spine* **2005**, *30*, E311–E314. [CrossRef] [PubMed]
89. Schulte, T. Acquired spondylolysis after implantation of a lumbar ProDisc II prosthesis: Case report and review of the literature. *Spine* **2007**, *32*, E645–E648. [CrossRef] [PubMed]
90. Pimenta, L.H.; Marchi, L.; Oliveira, L. Lumbar total disc replacement with a ball and socket metal on metal device: Up to 60-months follow-up. *Spine J.* **2012**, *12*, S104. [CrossRef]
91. Moghadas, P. Wear in metal-on-metal total disc arthroplasty. *Proc. Inst. Mech. Eng.* **2013**, *227*, 356–361. [CrossRef]
92. Dong-wook, K.; Lee, K.-Y.; Jun, Y.; Lee, S.J.; Park, C.K. Friction and wear characteristics of UHMWPE against Co-Cr alloy under the wide range of contact pressures in lumbar total disc. replacement. *Int. J. Precis. Eng. Manuf.* **2011**, *12*, 1111–1118. [CrossRef]
93. van Ooij, A.; Kurtz, S.M.; Stessels, F.; Noten, H.; van Rhijn, L. Polyethylene wear debris and long-term clinical failure of the CHARITE disc prosthesis: A study of 4 patients. *Spine* **2007**, *32*, 223–229. [CrossRef] [PubMed]

94. Lee, C. Development of a prosthetic intervertebral disc. *Spine* **1991**, *16*, S253–S255. [CrossRef] [PubMed]
95. Huang, R. Biomechanics of nonfusion implants. *Orthop. Clin. N. Am.* **2005**, *36*, 271–280. [CrossRef]

Article

Prediction of Wear in Crosslinked Polyethylene Unicompartmental Knee Arthroplasty

Jonathan Netter [1], Juan Hermida [1], Cesar Flores-Hernandez [1], Nikolai Steklov [1], Mark Kester [2] and Darryl D. D'Lima [1,*]

[1] Shiley Center for Orthopaedic Research and Education at Scripps Clinic, Scripps Health, 11025 North Torrey Pines Road, Suite 200, La Jolla, CA 92037, USA; jonnetter@gmail.com (J.N.); jhermida@gmail.com (J.H.); Flores-Hernandez.Cesar@scrippshealth.org (C.F.-H.); nick.steklov@gmail.com (N.S.)
[2] Stryker Orthopaedics, 325 Corporate Drive Court, Mahwah, NJ 07430, USA; mkesterphd85@gmail.com
* Author to whom correspondence should be addressed; ddlima@scripps.edu; Tel.: +1-858-332-0166; Fax: +1-858-332-0669.

Academic Editors: J. Philippe Kretzer and Amir Kamali
Received: 4 February 2015; Accepted: 24 April 2015; Published: 7 May 2015

Abstract: Wear-related complications remain a major issue after unicompartmental arthroplasty. We used a computational model to predict knee wear generated *in vitro* under diverse conditions. Inverse finite element analysis of 2 different total knee arthroplasty designs was used to determine wear factors of standard and highly crosslinked polyethylene by matching predicted wear rates to measured wear rates. The computed wear factor was used to predict wear in unicompartmental components. The articular surface design and kinematic conditions of the unicompartmental and tricompartmental designs were different. Predicted wear rate (1.77 mg/million cycles) was very close to experimental wear rate (1.84 mg/million cycles) after testing in an AMTI knee wear simulator. Finite element analysis can predict experimental wear and may reduce the cost and time of preclinical testing.

Keywords: unicompartmental arthroplasty; polyethylene wear; crosslinked polyethylene; finite element analysis; wear simulation

---

1. Introduction

Bearing wear has been a major reasons for complications after knee arthroplasty [1,2]. Although several biomaterials have been introduced to reduce wear in other joints, polyethylene continues to be the most popular material for the insert bearing in knee arthroplasty. Advances in crosslinking of polyethylene have been successful in dramatically reducing wear rates and osteolysis in hip arthroplasty [3]. The clinical success of crosslinked polyethylene in hip arthroplasty has led to its introduction in knee arthroplasty [4]. However, the inherent differences in hip and knee implant designs and differences in loading and kinematics do not permit direct extrapolation of the results of hip wear testing to knee wear testing.

Total knee arthroplasty (TKA) has an overall high success rate in terms of implant survival. Revision for wear is no longer as common as revision for infection, instability, and stiffness [5,6]. However, it is difficult to justify tricompartmental arthroplasty in young or middle-aged patients with unicompartmental disease. These patients are often more athletic and expect better function than the typical patient with tricompartmental disease. Long-term outcomes of high tibial osteotomy have not been consistent with survivorship that decreases to 75% at 10 years and 65% at 15 years [7,8]. Unicompartmental knee replacement can be an attractive alternative for this patient population [7,9,10].

Unicompartmental arthroplasty (UKA) preserves the rest of the joint and has been shown to better approximate normal knee kinematics than TKA [11]. However, survivorship after UKA is often lower than that in TKA [12,13]. Wear and polyethylene damage continue to be important factors affecting outcomes and have been implicated in up to 22% of revision surgeries after unicompartmental knee arthroplasty [14].

Knee wear simulation *in vitro* requires expensive equipment and several months of testing. Computational models of knee wear have been used to reproduce experimental wear. These wear factors are typically based on Archard's law, which assumes that wear is a function of local contact pressure and sliding distance [15]. Simulations of knee wear have been reported with a constant wear factor (first-generation Archard wear factors), with wear factors that vary by a function of the cross-shear (second-generation Archard wear factors), and by treating wear as a function of contact area (and sliding distance) rather than contact stress [16–19]. Common wear factors are those derived from pin-on-disk wear experiments. However, wear factors derived from pin-on-disk experiments, even those that vary by a function of the cross-shear, have not been shown to be accurate in predicting experimental or clinical wear rates [20]. In addition, very few of these models' predictions have been validated under diverse conditions [16,18].

We first used an inverse FEA approach to derive an "apparent" wear factor from our experimental knee wear simulations. The inverse FEA approach adjusts the wear factor used in the model until the predicted wear rate matches the experimental wear rate. Our objective was to determine if a wear factor computed for a specific combination of TKA bearing materials (polyethylene and cobalt-chrome (CoCr) alloy) and lubrication conditions could be used to predict wear in a different design (UKA) with the same materials but under different kinematic and loading conditions. We therefore used the FEA model to predict wear in UKA (using the wear factor obtained for TKA wear). The only change we made to the model was in the implant design (TKA to UKA) and in the kinematics and loading of the knee. The material properties of polyethylene, the algorithm to compute wear, and the wear factor were the same for both TKA and UKA models. Our objective was to predict UKA wear that closely matched experimental wear measurements and support the robustness of our model.

2. Materials and Methods

2.1. Experimental Wear Simulation of Tricompartmental Design

Gravimetric wear rates were obtained from previously reported experimental wear simulations of standard crosslinked polyethylene (Sigma® CR, DePuy, Warsaw, IN) and highly crosslinked polyethylene (Triathlon CR X3, Stryker, Mahwah, NJ, USA) [21,22]. We selected these 2 experimental studies based on differences in design, in bearing material, and in loading kinematic conditions. A computer model that could replicate both wear simulator studies would likely be more robust. One group ($n = 3$), comprised of moderately conforming cruciate-retaining inserts (Sigma® Curved, DePuy, Warsaw, IN, USA) subjected to 40 kGy sterilization irradiation (standard crosslinked). The second group ($n = 3$) consisted of moderately conforming cruciate-retaining inserts (Triathlon TKA) with UHMWPE inserts manufactured from compression molded GUR 1020 UHMWPE that were subjected to gamma irradiation to 30 kGy followed by an annealing process and then repeated 2 more times for a total dose of 90 kGy (highly crosslinked, X3®, Stryker). Femoral and tibial components for both groups were manufactured of CoCr alloy. Three implants from each of the 2 groups were tested on a 6-station displacement-controlled knee wear simulator (AMTI, Watertown, MA, USA).

Lubrication was provided by bovine serum (25%) supplemented with EDTA and sodium azide. Total protein concentration of the batch of bovine serum was 8.5 g/dL, with the albumin component being 3.2 g/dL (data provided by the supplier, Sigma-Aldrich, St. Louis, MO, USA). For the standard crosslinked tricompartmental design, the ISO-recommended wear testing protocol for displacement controlled knee wear simulators was used [23]. For the highly crosslinked tricompartmental design a more aggressive protocol was used, which involved a medial to lateral axial load distribution of 70:30 and a tibial axial rotation offset to simulate malaligned tibial components. Wear was monitored by visual inspection and by measuring weight loss at approximately 500,000 cycle intervals for a total of 5 million cycles. To correct for fluid absorption, implants of each group ($n = 3$) were soaked in bovine serum lubricant for the duration of the study and were subjected to the same dehydration and weighing protocol.

2.2. Computational Determination of Wear Factor

A computational model of the boundary conditions of the AMTI wear-testing machine was constructed in MSC.MARC (MSC Software, Santa Ana, CA, USA). Loading and motion conditions used in each of the experimental studies were replicated in the computer simulation. The axial load was force controlled, while flexion, anterior-posterior translation, and internal-external rotation were displacement controlled. The tibial tray was free to translate in the mediolateral direction and to rotate in varus-valgus (similar to the experimental protocol). CAD models of the implants (Figure 1) were meshed using HyperMesh (Altair Engineering, Inc., Troy, MI, USA). The polyethylene inserts were meshed with hexahedral element. Optimum mesh density was calculated in 2 stages. Predicted peak contact stresses and contact area in a simplified model of a spherical rigid body indenter with a radius of 14 mm (simulating a femoral condyle) contacting a flat "insert" (50 × 50 × 50 mm) with a linear elastic modulus of 700 MPa was compared to an analytic Hertzian contact solution [24]. Multiple loads (range, 100–1000 N) generating peak contact stresses of up to 70 MPa were simulated. Convergence of contact area measurement and peak contact stresses within 3% of the analytical solution was achieved with a mesh density using elements with a mean edge length of 1.5 mm. Next, convergence of peak contact stresses for the FEA model using the corresponding prosthetic component geometry were also obtained with a mean edge length of 1.5 mm (less than 1% change in peak contact stresses between mean element sizes of 1.5 mm and 1.0 mm). The inserts were modeled with a Young's Modulus of 700 MPa and Poisson's ratio of 0.43. While we have implemented nonlinear viscoelastic and elastoplastic material models in previous studies, we did not detect a significant difference in wear rates between linear and nonlinear material models [25,26]. The femoral and tibial metal components were simulated as rigid bodies. Contact was simulated between the femoral and the insert articular surfaces with a coefficient of friction of 0.2.

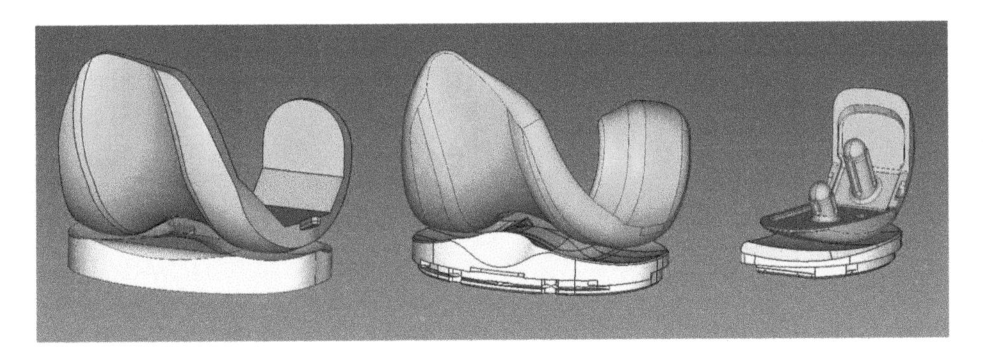

Figure 1. Finite element models were generated to simulate the experimental wear tests. **Left**: Sigma PFC standard crosslinked tricompartmental design; **Middle**: Stryker Triathlon X3 highly crosslinked tricompartmental design; **Right**: Stryker Triathlon PKR highly crosslinked design.

Contact area, contact pressure, and sliding distance were computed. Wear was predicted by using the Archard's classic law for mild wear.

$$\delta_{Wear} = k \sum_{i=1}^{n} \sigma_i |d_i| \Delta t_i \tag{1}$$

δ = material lost due to wear at each node; k = constant wear factor; σ = local contact stress; d = relative sliding distance; Δt = time increment; i = increment number; n = total number of increments for each cycle.

Each cycle was divided into 100 increments and wear was computed for each increment and summed over the entire cycle. The surface nodes affected by wear were moved in a direction normal to the articular surface based on the computed material loss at the end of every increment. The MSC.MARC solver adaptively adjusted interior nodes to maintain element quality. The simulation was then repeated in a step-wise manner and the wear was multiplied by the size of each step (50,000 cycles per step) to compute the total wear at the end of 5 million cycles. This update interval was shorter than those used in previously published FEA studies of TKA wear: 100,000 and 200,000 cycles [16]; 500,000 cycles [17]. For comparison with experimental data, computed volumetric wear was converted to gravimetric wear using polyethylene density of 0.97 mm^3/mg. The wear factors for the standard and highly crosslinked polyethylene were then determined by minimizing the difference between predicted wear and experimental wear rate using a global optimization tool (Isight, Dassault Systèmes, Providence, RI, USA). The wear factor obtained for highly crosslinked polyethylene was used to predict wear in Triathlon X3 PKR (partial knee replacement) components. To assess differences in kinematic patterns between experimental wear tests we also calculated crossing intensity using the method reported by Hamilton *et al.* [27].

2.3. *Experimental Wear Simulation of Unicompartmental Design*

PKR (unicompartmental) components were experimentally tested on the AMTI knee wear simulator. The implant materials (CoCr alloy and X3 polyethylene) and serum lubricant conditions were the same. However, the PKR is a unicompartmental design and the loading and kinematic conditions were significantly different from the tricompartmental wear simulation.

PKR inserts were mounted on top of the tibial tray and fixed by a press-fit mechanism supplemented with threaded bolts that allowed the inserts to be removed for cleaning and weighing without damaging the polyethylene. All components were mounted medial to the center of axial rotation of the knee station to simulate a medial compartmental knee arthroplasty.

Lubrication was the same as for the tricompartmental wear simulation. The ISO-recommended wear testing protocol for displacement controlled knee wear simulators was used. Since there was no lateral compartment, the entire axial load (representing 60% of the medial compartmental load) was transmitted through the unicompartmental components. Wear was monitored as described for the tricompartmental design. To correct for fluid absorption, additional inserts were soaked in the bovine serum lubricant for the duration of the study and were subjected to the same dehydration and weighing protocol.

3. Results

Wear factors generated for standard crosslinked and highly crosslinked polyethylene tricompartmental inserts are listed in Table 1. The simulated wear area visually approximated the experimental wear scars (Figure 2). Wear rates predicted by the finite element model for the PKR design were within 5% of those measured experimentally (Figure 3).

We analyzed the relative contribution of contact area, contact pressure (Figure 4), and sliding distance (Table 2). Instantaneous peak contact stresses averaged over the entire gait cycle and were 18.1 MPa for the standard crosslinked tricompartmental design, 26.8 MPa for the highly crosslinked tricompartmental design, and 31.1 MPa for the highly crosslinked PKR design (Figure 4).

Linear wear penetration was measured as the maximum depth of wear in the vertical direction and was computed for the standard crosslinked tricompartmental design, the highly crosslinked tricompartmental design, and the highly crosslinked PKR design. For comparison with clinical retrieval studies, the distribution of linear penetration was also calculated in the medial and lateral compartments (Table 3). Wear occurring during the stance phase was compared to that occurring during the swing phase and was expressed as a percentage of total wear (Table 4).

Table 1. Computed Wear Coefficients ($mm^3/N \times mm$).

Standard Crosslinked PE	Highly Crosslinked PE
1.70×10^{-10}	5.24×10^{-11}

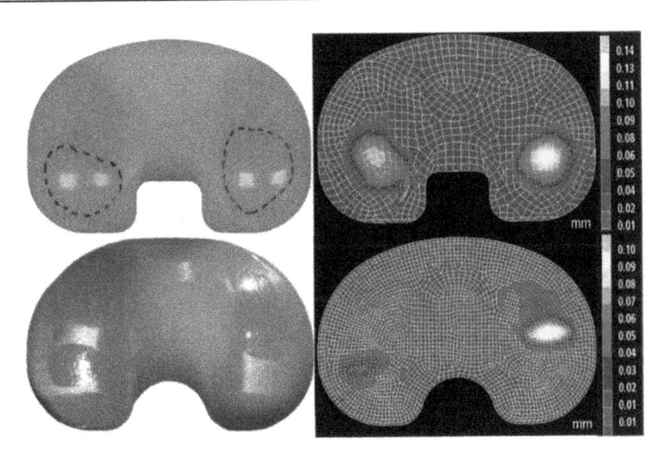

Figure 2. Photographs of experimental wear scar (**left**) were compared to predicted wear depth contour maps (**right**, mm) in the standard (**top**) and highly (**bottom**) crosslinked polyethylene (Sigma and Triathlon designs, respectively). The differences in location of the wear scars were due to the differences in kinematics between the two designs.

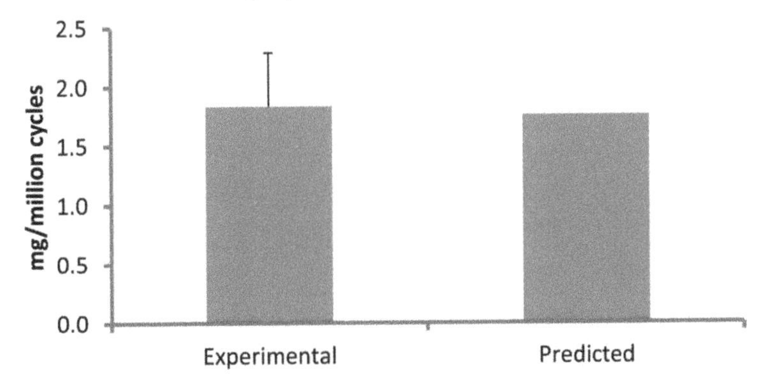

Figure 3. Computational predictions of wear rate (mg/million cycles) in the PKR design were very close to experimental measurements.

Table 2. Relative sliding distance (mm/cycle).

Triathlon TKR	Triathlon PKR
57.6	39.9

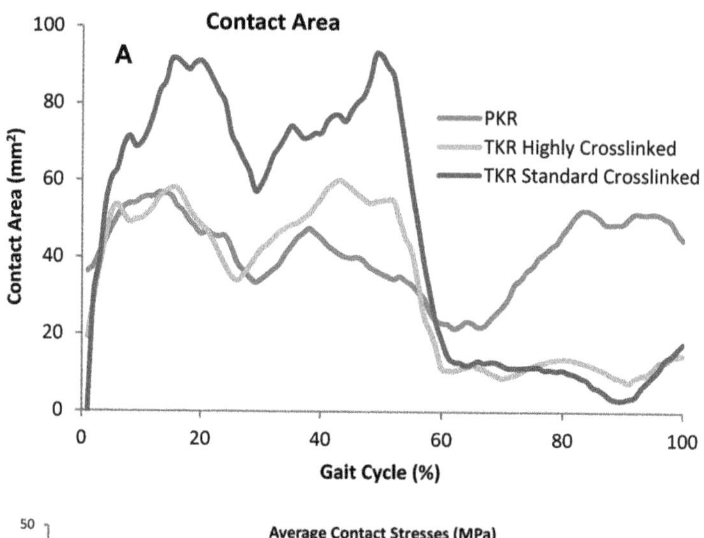

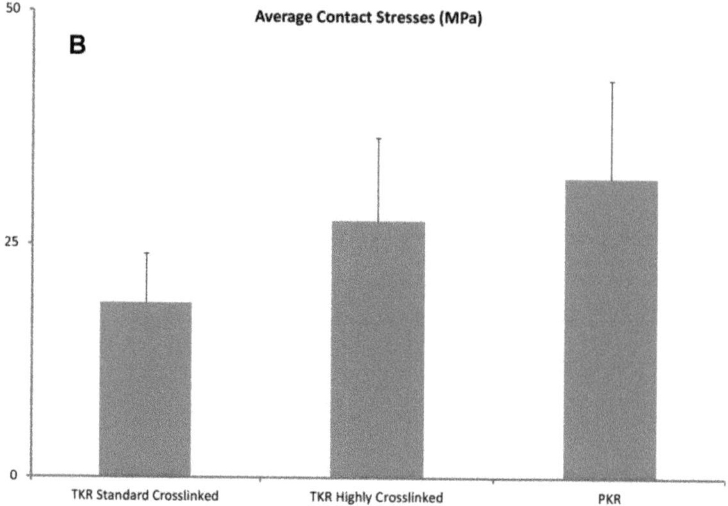

Figure 4. (**A**) Contact areas computed for each condition were plotted over the entire gait cycle. (**B**) Contact pressures were computed for all three designs and averaged across each cycle. Error bars denote standard deviation of area or peak contact stress over each cycle.

Table 3. Linear penetration rates (mm/million cycles).

Polyethylene type	M:L Load Distribution	Medial Penetration	Lateral Penetration
Highly Crosslinked TKA	70:30	0.025	0.018
Standard Crosslinked TKA	60:40	0.045	0.039
Standard Crosslinked TKA	68:32	0.054	0.029
Highly Crosslinked PKR	60:40	0.019	

Table 4. Wear in stance and swing phase (% of total wear).

Phase	Triathlon TKR	Triathlon PKR
Stance	56	62
Swing	44	38

4. Discussion

Measuring wear *in vivo* after knee arthroplasty is technically difficult. In vitro wear simulation has been validated as being reasonably predictive of clinical wear performance but requires expensive equipment and several months of test duration to complete the equivalent of five years of clinical performance [28]. We simulated wear rates under one set of implant design and kinematic conditions in a finite element model and used the model to predict wear in a different design under different kinematic conditions. Predicting wear under diverse conditions increased the validity of our model.

A finite element-based wear model generated the wear factor for crosslinked polyethylene (Triathlon X3) against the CoCr alloy using previously reported experimental wear measurements. The location of the wear scars on the experimentally tested inserts was visually congruent with the distribution of wear depth in the computer model. This wear factor was then used to predict wear for a unicompartmental design (Triathlon PKR) under significantly different loading and kinematic conditions. The model predicted wear rate (1.77 mg per million cycles) that was within 0.1 mg of the experimentally measured values (1.84 $\pm$ 0.45). We assessed the sensitivity of computed wear rate to the magnitude of wear factor and found that the wear rate changed by 0.18 mg/million cycles for each 10% difference in magnitude of wear factor. We also computed wear rates for two commonly reported constant wear factors: 2.64×10^{-10} and 1.07×10^{-9} mm^3/N-mm, which yielded wear rates of 9.07 and 37.38 mg/million cycles, respectively.

Validated computer models make it feasible to isolate the effects of major factors thought to contribute to wear, such as the wear coefficient for the bearing material pair, contact area and pressure between articulating surfaces, and the relative sliding distance. The Sigma Curved design has a higher tibiofemoral conformity, which is reflected in the greater contact area and lower contact pressure. Nevertheless, the crosslinked polyethylene wore substantially less, which emphasized the importance of significantly low coefficient of wear.

While linear penetration depth and volumetric wear are somewhat correlated, these outcome measures are sensitive to different factors. Small contact areas with high contact stresses generate higher linear penetration than volumetric wear, while large contact areas with low contact stresses can increase volumetric wear with lower linear penetration. This distinction can be important since a tendency toward greater linear penetration is more likely to result in localized damage to the insert (or even wear through); while a tendency toward greater volumetric wear (with low linear penetration) is more likely to generate greater volume of wear debris and can result in an osteolytic reaction.

One advantage of finite element models is the ability to identify wear rates during different phases of the gait cycle (such as stance phase wear *versus* swing phase wear rates), which is virtually impossible with experimental wear measurements. As anticipated, the predicted average wear rate was higher during the stance phase of the gait cycle. However, the wear rate during the swing phase was surprisingly high: 38% and 44% of total wear for the highly crosslinked PKR and TKR, respectively (Table 4). The ISO standard for knee wear simulation recommends an axial load of 168 N during the swing phase [23]. We have measured loads *in vivo* averaging 0.3 times body weight [29]. Despite the substantially lower loads during the swing phase, predicted wear was not insignificant. Crossing intensity can vary during the stance and swing phases; we calculated that average crossing intensity during swing phase was greater by 0.024. Our model assumed an apparent wear factor representative of the "average" amount of cross-shear, therefore the fraction of wear during the swing phase was likely to be even higher that our prediction, which was due to the small increase in crossing intensity.

Our analysis indicated this was a combination of greater sliding distance (due to the larger knee flexion arc during swing phase) and the moderate conformity of the articulating surfaces.

Clinically validated computer models are extremely valuable for efficient preclinical testing of wear performance and for supporting design development. An analysis of retrieved knee non-crosslinked inserts revealed linear penetration rates of 0.054 and 0.029 mm/year for the medial and lateral compartments, respectively [30]. Our predictions for the standard crosslinked tricompartmental design (0.045 medial; 0.039 lateral) were within 1 standard deviation of the reported clinical penetration. Our predictions were based on replicating experimental wear conditions in which 60% of axial load was distributed to the medial compartment and 40% to the lateral compartment. When we reran the wear simulations with a medial to lateral load distribution of 68:32% (Table 3), the predicted linear penetration was almost identical to the clinical measurements. This result is consistent with finding that 73% of the first peak and 65% of the second peak of *in vivo* knee forces are transmitted through the medial compartment [31]. A more recent retrieval analysis of the same design (Sigma Curved CR) that we studied revealed an average wear rate (combined medial and lateral) of 0.08 mm/year [32]. This result was higher than our prediction. One likely explanation was the presence of creep in the retrievals that was ignored in our simulations. Also, we assumed 1 million cycles to be equivalent to one year of clinical activity, while pedometer measurements have revealed that active patients walked 3.5 times that average number of steps per day [33]. Wear rates have been even higher in retrievals of older designs with flat-on-flat tibiofemoral conformity, which has been shown to wear more than the curved-on-curved designs such as the ones tested in the present study [34,35].

One limitation of our model was that we used a constant wear factor that did not change with contact stress or with sliding direction. It has been shown that the wear factor increases with the degree of multidirectional cross-shear [36,37]. Reports on the accuracy of predictions of wear rates based on the use of various wear factors are not consistent. A computationally efficient rigid-body-based approach supplemented with discretized springs for contact analysis found that a simple Archard wear constant was not able to predict experimental wear across a broad number of implant designs and conditions, while wear factors that changed with the degree of cross-shear were more accurate [18]. On the other hand, an explicit FEA approach predicted experimental knee wear reasonably well using a constant wear factor averaged from the literature [17]; wear factors derived from pin-on-disk experiments that vary by a function of the cross-shear have not been shown to be accurate in predicting experimental or clinical wear rates [20]. Another approach taken was to assume that wear is a function of the contact area (instead of stress) and sliding distance and uses a factor that varied by cross-shear [38]. However, that approach underestimated experimental wear in total hip arthroplasty by 40%. Using the same method in TKA also generated predictions of wear rates of similar accuracy as for total hip arthroplasty [16]. Another limitation of our model was that suboptimal positioning of implants was not simulated. We have previously shown that implant alignment has a significant effect on wear in retrievals [39].

The intensity of multi-directional cross-shear has been defined as a ratio of the perpendicular sliding distance (A) to the sum of the sliding distance in the principal direction of motion (B) and A *i.e.*, A/(A+B) [40]; in terms of the incremental deviation of the slip direction relative to the dominant direction of orientation [27]; or as the ratio of the frictional work in the direction perpendicular to the direction of the principal polymer orientation to the total frictional work [41]. Pin-on-disk wear experiments indicate that the wear factor changes very slowly above a cross-shear ratio of 0.04 [41]. Finally, crosslinked polyethylene is practically insensitive to cross-shear ratios above 0.04 presumably because the cross-links prevent molecular reorientation to the principal direction of motion [41]. In the knee, during a gait cycle, crossing intensity is lower than in hip or intervertebral disk replacements (<0.08) [27]. We also used the method reported by Hamilton *et al.* to calculate crossing intensity. In our study, the difference in average crossing intensity between the tricompartmental and unicompartmental designs was 0.004. Collectively, these results indicate that within a range of cross-shear ratios, the wear prediction may be approximated with an "apparent" constant wear factor that approximates an average of the crossing motion within the range of that found during normal walking.

We therefore chose not to include the effect of multi-directional cross-shear in the interest of generating a simpler model that captured wear performance during walking with acceptable accuracy. Our assumption was that a constant Archard wear factor derived from experimental wear combines the contributions of cross-shear, friction, and slip velocity to wear of a given bearing couple. Despite these simplifications, our model was able to predict wear in a unicompartmental design with different kinematics within reasonable accuracy.

Computational models of knee wear have been shown to reproduce the wear generated *in vitro*. However, these models have not been validated under diverse conditions. A validated computer model that can predict wear in different knee arthroplasty designs and under different loading conditions is extremely valuable. Knee wear simulator results are commonly used to predict implant performance in patients. Wear testing for 5 million cycles on knee wear simulators typically takes about three months, while the time required for constructing and solving one computational wear simulation was less than a week. This method allows one to estimate the wear performance of a new device even before expensive components are manufactured. Predicting wear in validated and efficient computational models now permits design evaluation *in silico* and may even be used to optimize design parameters.

5. Conclusions

We successfully predicted wear in partial knee replacement using wear factors derived from TKA experiments using an inverse FEA approach. Previously, wear factors derived from pin-on-disk experiments were not very successful in accurately predicting wear. Furthermore, validating the model under diverse kinematic and loading conditions increased the robustness of our model. Minimizing wear rate remains essential to implant survivorship. This approach may be used to predict wear in existing implant designs, and moreover, may also be used to optimize design parameters of new devices. Our method can significantly accelerate wear evaluation.

Author Contributions: Darryl D. D'Lima and Mark Kester conceived and designed the experiments; Juan Hermida, Nikolai Steklov, and Jonathan Netter performed the experiments; Jonathan Netter and Cesar Flores-Hernandez generated the computational models; Jonathan Netter and Darryl D. D'Lima analyzed the data; Mark Kester contributed implant components; Jonathan Netter and Darryl D. D'Lima wrote the paper.

Conflicts of Interest: One author was an employee of Stryker Orthopaedics. The other authors declare no conflicts of interest.

References

1. Sharkey, P.F.; Hozack, W.J.; Rothman, R.H.; Shastri, S.; Jacoby, S.M. Insall Award paper. Why are total knee arthroplasties failing today? *Clin. Orthop. Relat. Res.* **2002**, *404*, 7–13. [CrossRef] [PubMed]
2. Callaghan, J.J.; O'Rourke M, R.; Saleh, K.J. Why knees fail: Lessons learned. *J. Arthroplast.* **2004**, *19*, 31–34. [CrossRef]
3. Kurtz, S.M.; Gawel, H.A.; Patel, J.D. History and systematic review of wear and osteolysis outcomes for first-generation highly crosslinked polyethylene. *Clin. Orthop. Relat. Res.* **2011**, *469*, 2262–2277. [CrossRef] [PubMed]
4. Hodrick, J.T.; Severson, E.P.; McAlister, D.S.; Dahl, B.; Hofmann, A.A. Highly crosslinked polyethylene is safe for use in total knee arthroplasty. *Clin. Orthop. Relat. Res.* **2008**, *466*, 2806–2812. [CrossRef] [PubMed]
5. Sharkey, P.F.; Lichstein, P.M.; Shen, C.; Tokarski, A.T.; Parvizi, J. Why are total knee arthroplasties failing today—Has anything changed after 10 years? *J. Arthroplast.* **2014**, *29*, 1774–1778. [CrossRef]
6. Le, D.H.; Goodman, S.B.; Maloney, W.J.; Huddleston, J.I. Current modes of failure in TKA: Infection, instability, and stiffness predominate. *Clin. Orthop. Relat. Res.* **2014**, *472*, 2197–2200. [CrossRef] [PubMed]
7. Weale, A.E.; Newman, J.H. Unicompartmental arthroplasty and high tibial osteotomy for osteoarthrosis of the knee. A comparative study with a 12- to 17-year follow-up period. *Clin. Orthop. Relat. Res.* **1994**, *302*, 134–137. [PubMed]

8. Hui, C.; Salmon, L.J.; Kok, A.; Williams, H.A.; Hockers, N.; van der Tempel, W.M.; Chana, R.; Pinczewski, L.A. Long-term survival of high tibial osteotomy for medial compartment osteoarthritis of the knee. *Am. J. Sports Med.* **2011**, *39*, 64–70. [CrossRef] [PubMed]
9. Broughton, N.S.; Newman, J.H.; Baily, R.A. Unicompartmental replacement and high tibial osteotomy for osteoarthritis of the knee. A comparative study after 5–10 years' follow-up. *J. Bone Joint Surg. Br.* **1986**, *68*, 447–452. [PubMed]
10. Pandit, H.; Jenkins, C.; Gill, H.S.; Barker, K.; Dodd, C.A.; Murray, D.W. Minimally invasive Oxford phase 3 unicompartmental knee replacement: Results of 1000 cases. *J. Bone Joint Surg. Br.* **2011**, *93*, 198–204. [CrossRef] [PubMed]
11. Patil, S.; Colwell, C.W., Jr.; Ezzet, K.A.; D'Lima, D.D. Can normal knee kinematics be restored with unicompartmental knee replacement? *J. Bone Joint Surg. Am.* **2005**, *87*, 332–338. [CrossRef] [PubMed]
12. Parratte, S.; Pauly, V.; Aubaniac, J.M.; Argenson, J.N. No Long-term Difference Between Fixed and Mobile Medial Unicompartmental Arthroplasty. *Clin. Orthop. Relat. Res.* **2012**, *470*, 61–68. [CrossRef] [PubMed]
13. Kuipers, B.M.; Kollen, B.J.; Bots, P.C.; Burger, B.J.; van Raay, J.J.; Tulp, N.J.; Verheyen, C.C. Factors associated with reduced early survival in the Oxford phase III medial unicompartment knee replacement. *Knee* **2010**, *17*, 48–52. [CrossRef] [PubMed]
14. Lidgren, L.; Robertsson, O.; W-Dahl, A. *The Swedish Knee Arthroplasty Register Annual Report*. 2009. Available online: www.ort.lu.se/knee/ (accessed on 25 April 2012).
15. Archard, J.F. Contact and rubbing of flat surfaces. *J. Appl. Phys.* **1953**, *24*, 981–988. [CrossRef]
16. Abdelgaied, A.; Liu, F.; Brockett, C.; Jennings, L.; Fisher, J.; Jin, Z. Computational wear prediction of artificial knee joints based on a new wear law and formulation. *J. Biomech.* **2011**, *44*, 1108–1116. [CrossRef] [PubMed]
17. Knight, L.A.; Pal, S.; Coleman, J.C.; Bronson, F.; Haider, H.; Levine, D.L.; Taylor, M.; Rullkoetter, P.J. Comparison of long-term numerical and experimental total knee replacement wear during simulated gait loading. *J. Biomech.* **2007**, *40*, 1550–1558. [CrossRef] [PubMed]
18. Strickland, M.A.; Taylor, M. *In-silico* wear prediction for knee replacements—Methodology and corroboration. *J. Biomech.* **2009**, *42*, 1469–1474. [CrossRef] [PubMed]
19. Zhao, D.; Sakoda, H.; Sawyer, W.G.; Banks, S.A.; Fregly, B.J. Predicting knee replacement damage in a simulator machine using a computational model with a consistent wear factor. *J. Biomech. Eng.* **2008**, *130*. [CrossRef]
20. Kang, L.; Galvin, A.L.; Fisher, J.; Jin, Z. Enhanced computational prediction of polyethylene wear in hip joints by incorporating cross-shear and contact pressure in additional to load and sliding distance: Effect of head diameter. *J. Biomech.* **2009**, *42*, 912–918. [CrossRef] [PubMed]
21. Hermida, J.; Patil, S.; Chen, P.C.; McNulty, D.; Swope, S.; Colwell, C.W.; D'Lima, D.D. Total Knee Arthroplasty Design Affects Polyethylene Wear. In Proceedings of the Annual Meeting AAOS, New Orleans, LA, USA, 5–9 February 2003; Volume 4, p. 398.
22. Hermida, J.C.; Fischler, A.; Colwell, C.W., Jr.; D'Lima, D.D. The effect of oxidative aging on the wear performance of highly crosslinked polyethylene knee inserts under conditions of severe malalignment. *J. Orthop. Res.* **2008**, *26*, 1585–1590. [CrossRef] [PubMed]
23. *Standard Number 14243-3:2014: Implants for Surgery—Wear of Total Knee Joint Prostheses—Part 3: Loading and Displacement Parameters for Wear-Testing Machines with Displacement Control and Corresponding Environmental Conditions for Test*; International Organization for Standardization (ISO): Geneva, Switzerland, 2014.
24. Fischer-Cripps, A.C. *Introduction to Contact Mechanics*; Springer-Verlag: New York, NY, USA, 2000.
25. D'Lima, D.D.; Chen, P.C.; Colwell, C.W., Jr. Polyethylene contact stresses, articular congruity, and knee alignment. *Clin. Orthop. Relat. Res.* **2001**, *392*, 232–238. [CrossRef] [PubMed]
26. D'Lima, D.D.; Steklov, N.; Fregly, B.J.; Banks, S.; Colwell, C.W., Jr. *In vivo* contact stresses during activities of daily living after knee arthroplasty. *J. Orthop. Res.* **2008**, *26*, 1549–1555. [CrossRef] [PubMed]
27. Hamilton, M.A.; Sucec, M.C.; Fregly, B.J.; Banks, S.A.; Sawyer, W.G. Quantifying multidirectional sliding motions in total knee replacements. *J. Tribol.* **2005**, *127*, 280–286. [CrossRef]
28. McKellop, H.A.; D'Lima, D. How have wear testing and joint simulator studies helped to discriminate among materials and designs? *J. Am. Acad. Orthop. Surg.* **2008**, *16*, S111–S119. [PubMed]
29. D'Lima, D.D.; Patil, S.; Steklov, N.; Slamin, J.E.; Colwell, C.W., Jr. Tibial forces measured *in vivo* after total knee arthroplasty. *J. Arthroplast.* **2006**, *21*, 255–262. [CrossRef]

30. Engh, G.A.; Zimmerman, R.L.; Parks, N.L.; Engh, C.A. Analysis of wear in retrieved mobile and fixed bearing knee inserts. *J. Arthroplast.* **2009**, *24*, 28–32. [CrossRef]
31. Halder, A.; Kutzner, I.; Graichen, F.; Heinlein, B.; Beier, A.; Bergmann, G. Influence of limb alignment on mediolateral loading in total knee replacement: *In vivo* measurements in five patients. *J. Bone Joint Surg. Am.* **2012**, *94*, 1023–1029. [CrossRef] [PubMed]
32. Berry, D.J.; Currier, J.H.; Mayor, M.B.; Collier, J.P. Knee wear measured in retrievals: A polished tray reduces insert wear. *Clin. Orthop. Relat. Res.* **2012**, *470*, 1860–1868. [CrossRef] [PubMed]
33. Schmalzried, T.P.; Szuszczewicz, E.S.; Northfield, M.R.; Akizuki, K.H.; Frankel, R.E.; Belcher, G.; Amstutz, H.C. Quantitative assessment of walking activity after total hip or knee replacement. *J. Bone Joint Surg. Am.* **1998**, *80*, 54–59. [CrossRef] [PubMed]
34. Lavernia, C.J.; Sierra, R.J.; Hungerford, D.S.; Krackow, K. Activity level and wear in total knee arthroplasty: A study of autopsy retrieved specimens. *J. Arthroplast.* **2001**, *16*, 446–453. [CrossRef]
35. Benjamin, J.; Szivek, J.; Dersam, G.; Persselin, S.; Johnson, R. Linear and volumetric wear of tibial inserts in posterior cruciate-retaining knee arthroplasties. *Clin. Orthop. Relat. Res.* **2001**, *392*, 131–138. [CrossRef] [PubMed]
36. Bragdon, C.R.; O'Connor, D.O.; Lowenstein, J.D.; Jasty, M.; Syniuta, W.D. The importance of multidirectional motion on the wear of polyethylene. *Proc. Inst. Mech. Eng. H* **1996**, *210*, 157–165. [CrossRef] [PubMed]
37. Wang, A.; Stark, C.; Dumbleton, J.H. Mechanistic and morphological origins of ultra-high molecular weight polyethylene wear debris in total joint replacement prostheses. *Proc. Inst. Mech. Eng. H* **1996**, *210*, 141–155. [CrossRef] [PubMed]
38. Liu, F.; Galvin, A.; Jin, Z.; Fisher, J. A new formulation for the prediction of polyethylene wear in artificial hip joints. *Proc. Inst. Mech. Eng. H* **2011**, *225*, 16–24. [CrossRef] [PubMed]
39. Srivastava, A.; Lee, G.Y.; Steklov, N.; Colwell, C.W., Jr.; Ezzet, K.A.; D'Lima, D.D. Effect of tibial component varus on wear in total knee arthroplasty. *Knee* **2012**, *19*, 560–563. [CrossRef] [PubMed]
40. Turell, M.; Wang, A.; Bellare, A. Quantification of the effect of cross-path motion on the wear rate of ultra-high molecular weight polyethylene. *Wear* **2003**, *255*, 1034–1039. [CrossRef]
41. Kang, L.; Galvin, A.L.; Brown, T.D.; Jin, Z.; Fisher, J. Quantification of the effect of cross-shear on the wear of conventional and highly cross-linked UHMWPE. *J. Biomech.* **2008**, *41*, 340–346. [CrossRef] [PubMed]

 *lubricants*

Review

The Synovial Lining and Synovial Fluid Properties after Joint Arthroplasty

Michael Shang Kung [1], John Markantonis [1], Scott D. Nelson [2] and Patricia Campbell [1,*]

[1] J. Vernon Luck Sr. M.D. Orthopaedic Research Center, Orthopaedic Institute for Children/UCLA, Los Angeles, CA 90007, USA; kungms@gmail.com (M.S.K.); do17.john.markantonis@nv.touro.edu (J.M.)
[2] Departments of Pathology, and Orthopaedic Surgery, UCLA Santa Monica/Orthopaedic Hospital, Santa Monica, CA 90404, USA; sdnelson@mednet.ucla.edu
* Author to whom correspondence should be addressed; pcampbell@mednet.ucla.edu; Tel.: +1-213-742-1134; Fax: +1-213-742-3253.

Academic Editors: J. Philippe Kretzer and Amir Kamali
Received: 18 February 2015; Accepted: 5 May 2015; Published: 18 May 2015

Abstract: The lubrication of the cartilaginous structures in human joints is provided by a fluid from a specialized layer of cells at the surface of a delicate tissue called the synovial lining. Little is known about the characteristics of the fluids produced after a joint arthroplasty procedure. A literature review was carried out to identify papers that characterized the synovial lining and the synovial fluids formed after total hip or knee arthroplasty. Five papers about synovial lining histology and six papers about the lubricating properties of the fluids were identified. The cells making up the re-formed synovial lining, as well as the lining of interface membranes, were similar to the typical Type A and B synoviocytes of normal joints. The synovial fluids around joint replacement devices were typically lower in viscosity than pre-arthroplasty fluids but the protein concentration and phospholipid concentrations tended to be comparable, suggesting that the lining tissue function was preserved after arthroplasty. The widespread, long-term success of joint arthroplasty suggests that the lubricant formed from implanted joint synovium is adequate for good clinical performance in the majority of joints. The role the fluid plays in component wear or failure is a topic for future study.

Keywords: synovial lining; synovial fluid; lubrication; histomorphological characterization; arthroplasty; implant; systematic review

---

1. Introduction

1.1. Synovial Tissue

An appreciation of the microanatomy of the lining tissues of the human articular joints is important to the understanding of the clinical performance of joint replacement devices used to restore function to degenerative and painful joints. For the normal cartilage-capped articular joint, lubrication of the bony, cartilaginous, ligamentous, and fibrous tissue structures is provided by a fluid produced from a specialized layer of cells found at the surface of a delicate tissue called the synovial lining (from ovum or egg, because of the similarity with egg white). This synovial tissue is composed of specialized layers of cells starting at the outermost (joint-facing) level with the intima, which overlies a vascular subintimal layer, and is supported by a fibrous stroma forming the joint capsule [1–3]. Of note, there is a wide range of variability in the thickness of the intimal and subintimal layers, the degree of continuity and cellularity of the intima and the presence of folds and villi (finger-like projections) depending upon numerous factors including the health of the joint, the age of the host, and the location of the sample within the joint.

The synovial tissue has been described as "controlling the environment of the joint" [4]. It does this by participating in immune reactions to bacteria, phagocytosis (removal of detritus material), lubrication, and cartilage nutrition [1,5]. In young, healthy synovium, the 1–2 cell thick intimal layer is composed of macrophage-like type A cells which are mainly phagocytic and absorptive in nature, and fibroblast-like type B cells which are secretory in nature [1,2,6–11]. This intimal layer shows low degrees of cell division, suggesting that many of the lining cells migrate to the synovial surface from underlying blood vessels and that they are bone-marrow-derived cells [12]. The synovial tissue is able to respond quickly to injury or disuse. The changes can include thickening of the intimal layer, increased vascularity and inflammatory cell infiltration of the subsynovial layer, and the formation of multiple finger-like projections (villi) into the joint. The synovial tissue of an aged individual with osteoarthritis will look very different grossly and microscopically from that of a healthy young individual.

The surgical removal of chronically inflamed synovial tissue from arthritic, tuberculoid or traumatized joints was once a common orthopaedic procedure; fortunately, the synovial tissue can regenerate [13,14]. It is this regenerative capacity that allows implanted artificial joints to be lubricated despite surgical removal or damage to the synovial tissues. The microanatomy of the re-formed joint lining tissue and the quality of the lubricating fluid that it produces are the subjects of the present review.

Following joint replacement surgery, in addition to the regeneration of the lining of the joint capsule, membranes that form between the bone and the cement, or bone and implant (known as the interface) often show a synovium-like microstructure (so-called pseudo-synovial linings) [15–19]. It has been suggested that motion at the interface induces the production of this specialized tissue because it was observed in association with loose implants [20,21]. More recently, it has been shown that one of the constituents of synovial fluid, hyaluronan (also referred as hyaluronic acid), can induce the synovial-like interface membrane features [22]. A synovial-like membrane can be produced in the rat or mouse dorsal air pouch by disrupting the subcutaneous connective tissue by repeated injection of air [23]. The resulting cavity develops a lining structure as early as six days later which has many features of a synovial lining, including the production of fluid, and this has become a useful experimental model of the periprosthetic tissue [23–25].

1.2. Synovial Fluid

The lubrication for the natural joint is provided by a small amount of viscous fluid consisting of mainly hyaluronic acid, proteins and other factors such as lubricin and surface-active phospholipid (SAPL) that allow for effective gliding of the joints [26]. As with the tissue, the characteristics and properties of the fluid vary considerably as the result of joint health, injury status and other host factors such as inflammation and hypersensitivity. Injury to the joint may result in an abundance of fluid (a joint effusion), distension of the tissues and interruption of the normal functions of the synovial tissue [13]. Under normal biological conditions, the coefficient of friction is only 0.001 to 0.006 and wear is nearly zero because of the special tribologic properties of cartilage and synovial fluid [27]. Boundary lubrication is thought to be provided by surface-active phospholipid (SAPL) complexed with lubricin and hyaluronic acid, which allows for load-bearing capacities for the joint [28,29]. Lubricin, a component found to be plentiful in synovial fluid, is thought to be the lubricating molecule that facilitates the gliding motion of joints [30]. Lubricin may be a carrier of surface-active phospholipids because the SAPL are super hydrophobic and adsorb to the joint surfaces. Hyaluronic acid (HA), found abundantly in synovial fluid, while providing a lubrication effect, does not provide the load-bearing capabilities for joints [31]. Instead, the HA portion is thought to maintain the viscosity of the synovial fluid [32–34] and facilitate the formation of lipid multilayers on the lubricating surface by complexing with phospholipids, providing lubrication at the joint [35]. The purpose of this review was to examine the state of knowledge regarding the changes that occur in the synovial lining and lubrication properties of the surrounding fluid after implantation of a total joint prosthesis. This review will (1) summarize the characteristics of the synovial lining tissue histomorphology and (2) the composition of synovial fluid from published studies.

2. Methods

2.1. Synovial Lining Papers

A review of PubMed- and Embase-indexed studies up to November 2014 was carried out to identify studies regarding the histological characterization of the synovial lining of joints following total hip or knee replacement (hereafter called implanted tissue). This electronic search was supplemented by a manual search of the reference lists of the included papers. Conference proceedings, single case studies, and reviews were not included in the search and only studies in English language were included. Search strings included various combinations of the terms "synovium OR synovial lining OR synovial-like membrane OR interface membrane OR synovial lining cells OR synovitis OR intima, arthroplasty OR implant OR implanted, histological OR histology, and morphology OR morphological." Articles of interest were selected if detailed histomorphological characterization of the synovial lining or the interface membrane was performed and described. Articles were excluded for the following reasons: *in vitro* studies, animal studies, studies involving infection or rheumatoid arthritis exclusively, non-retrieval studies, and non-arthroplasty related studies. Articles that focused on the changes of the subintimal layers due to wear debris, but without describing the intima, were excluded from final selection. The resulting potential studies were reviewed, summarized, and critically evaluated (Figure 1).

Five articles were selected which reported the histological characterization of a synovial lining or a synovial-like membrane from implanted tissue and evaluated various bearing material combinations. Tissue specimens from a total of 173 patients were histologically examined and the morphology of the intimal layer was characterized in detail. Immunohistochemistry was performed using monoclonal and polyclonal antibodies to identify the cell subsets or the presence of prostaglandins, fibronectin, and other factors produced by the lining cells. All papers involved retrospective studies from implanted tissue. A summary of the overall findings of each selected article are presented in Tables 1 and 2.

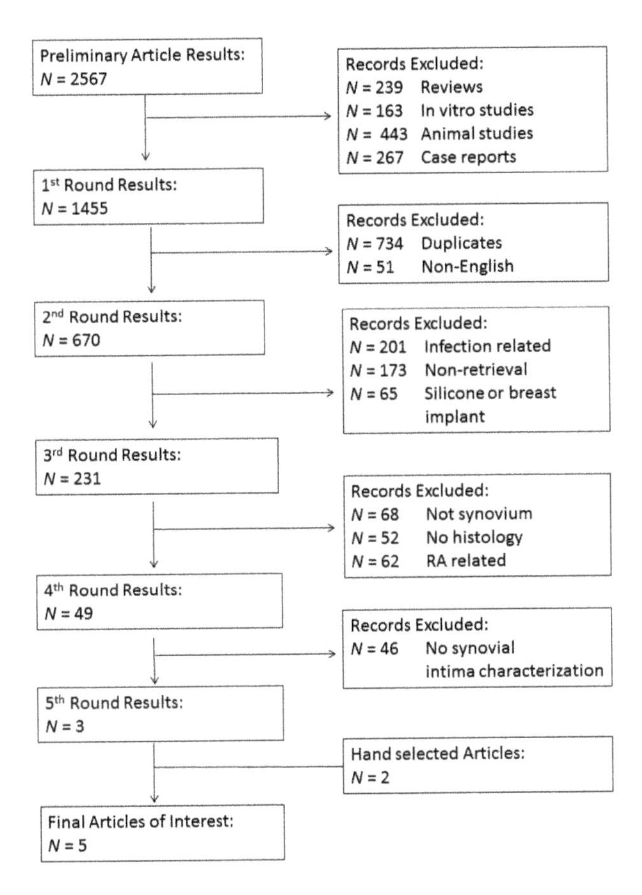

Figure 1. Flowchart of synovial lining article selection.

Table 1. Overall general findings for articles selected describing synovial lining characterization.

Author	Bearing Type [1]	Patients	Characterization of Synovial-Like Membrane			Main Synovium Findings
			Layer Thickness	Synovial Lining Cells	Presence of Particulate Debris in Synovial Lining [2]	
Goldring *et al.* (1983) [17]	Metal on Polyethylene	$N = 20$	cells	• Polygonal lining cells • Large stellate cells with multiple thin dendritic processes • some scattered giant cells	0	Membrane at cement-bone interface with histological and histochemical characteristics of normal synovium found in patients with implant loosening.
Goldring *et al.* (1986) [16]	Metal on Polyethylene	$N = 41$	1–2 cells	• Large polygonal cells with eccentric nuclei • Few scattered multinucleated giant cells	1	Study confirmed formation of synovial-like lining at bone-cement interface taken from patients with loosened THA components
Lennox *et al.* (1987) [19]	Metal on Polyethylene; Ceramic on Polyethylene	$N = 61$	1–3 cells	• Large, rounded or teardrop-shaped cells • Large Dendritic-shaped cells with long cellular processes extending toward surface	1	Cemented, press-fit, and biologic ingrowth prostheses showed similar formation of pseudosynovial lining at the implant-bone interface membrane.
Lalor and Revell (1993) [18]	Metal on Polyethylene and Polyethylene on Delrin THA and TKA	$N = 29$, 23 hip; 6 knee	1–10 cells	• OKM1+/OKM5+ macrophage-like cells • MAB67+ fibroblast-like cells	1	The newly formed bone-implant interface membrane closely resembled true synovium and contained macrophage-like type A cells and fibroblast type B cells, but not necessarily always in distinct layers
Burkandt *et al.* (2011) [15]	Metal on Polyethylene THA; Metal on Metal HRA	$N = 22$, 10 with synovitis; 12 with arthroplasty.	1–5 cells	• CD163+ macrophage-like cells • D2-40+ fibroblast-like cells	1	Tissues from patients revised due to suggested metal hypersensitivity showed increased proliferation of synovial lining cell layer similar to cases with rheumatoid arthritis and high-grade synovitis, with 2 patients showing paucicellular synovial membrane covered by a fibrinous exudate.

[1] Specimens from revised hip arthroplasty unless otherwise noted and this column gives the bearing combinations; [2] the relative amounts of debris noted histologically in excised synovial lining layer: 0 = none, 1 = rare, 2 = minimal, 3 = moderate, 4 = severe; THA = total hip arthroplasty, HRA = hip resurfacing arthroplasty, TKA = Total Knee Arthroplasty; Macrophage markers: OKM1, OKM5, CD163+; Fibroblast markers: MAB67, D2-406.

Table 2. Overall general findings for articles selected describing synovial fluid properties.

Author	Arthroplasty	Experiment	Key Findings
Costa *et al.* (2001) [36]	10 UHMWPE hip implants	Mass spectrophotometry and FTIR were performed after cyclohexane extraction for adsorbed products on the liners.	Methyl esters of hexadecanoic acid, octadecanoic acid, squalene, and of cholesterol were found in the extracts as well as a protein-like material at the surface.
Mazzucco *et al.* (2002) [37]	58 index TKA; 19 revision TKA; 2 effused previous TKA	Sufficient SF samples were obtained from 36 index TKA, 14 revised TKA, and 2 effused previous TKA for flow property examination.	SF from revision TKA tended to have lower viscosity than that from index TKA. The difference was found to not be statistically significant.
Mazzucco *et al.* (2004) [38]	77 index TKA; 20 revised TKA; 3 effused previous TKA	SF from 24 index TKA and 7 revised TKA had their composition of protein, phospholipids and HA determined and correlated.	Protein and phospholipids were found to have a positive correlation in regards to each other. Protein and phospholipids were found to have a negative correlation with HA.
Gale *et al.* (2007) [39]	38 Metal on Polyethylene THA; 2 Metal on Polyethylene TKA	The bearing surfaces of the implants were rinsed and analyzed by HPLC for phospholipids.	8 species of phosphatidylcholine were identified. 3 species of unsaturated phosphatidylcholine predominated; PLPC, POPC, and SLPC.
Bergmann *et al.* Part 1 (2001) [40]	Type 1 telemeterized cemented PE cup, 1 temperature measurement at neck; Type 2 telemeterized non-cemented, AC head, PE or AC cup, titanium shaft	Patients were monitored doing various physical activities and the temperatures inside their telemeterized implant were recorded.	The highest peak temperature were observed in the head of the implant and reached as high as 43.1 °C, greater than what is believed to affect the synovial fluids lubrication ability.
Bergmann *et al.* Part 2 (2001) [41]	(See above)	Data from Bergmann *et al.* 2001 part 1 was used to generate a finite element model to calculate the steady-state within the implant during walking.	The model shows that if the cup of an implant is made of a material with good conductivity, heat will be transferred away from the synovial fluid, capsule and stem towards the acetabular bone.

THA = total hip arthroplasty, TKA = Total Knee Arthroplasty, MS = Mass spectrometry, FTIR = Fourier transform infrared spectroscopy, AC = aluminum oxide ceramic, PE = polyethylene, SF = synovial fluid, (PLPC) = Palmitoyl-linoleoyl-phosphatidylcholine, POPC = Palmitoyl-oleoyl-phosphatidylcholine, SLPC = Stearoyl-linoleoyl-phosphatidylcholine (SLPC).

2.2. *Synovial Lubrication Papers*

A review of PubMed studies from the time period of 2000 through 2014 was performed to find papers relevant to the lubrication characteristics of synovial fluid (SF) after joint arthroplasty. Search strings included "Synovial Fluid AND Prosthetic (136 results), Synovial Fluid AND Arthroplasty (391 results), Synovial Fluid AND Implant (91 results) and Synovial Fluid AND Revision (132 results)" for a total of 750 articles. The article titles and abstracts were examined and papers were excluded for the following reasons: SF analysis of biomarkers for periprosthetic infections, ion/cytokine/particle/cell levels, arthritic joint analysis before arthroplasty only, *in vitro* studies, pain only, wear/friction only, animal studies, case studies, literature reviews, and surgical procedure and imaging studies only (Figure 2).

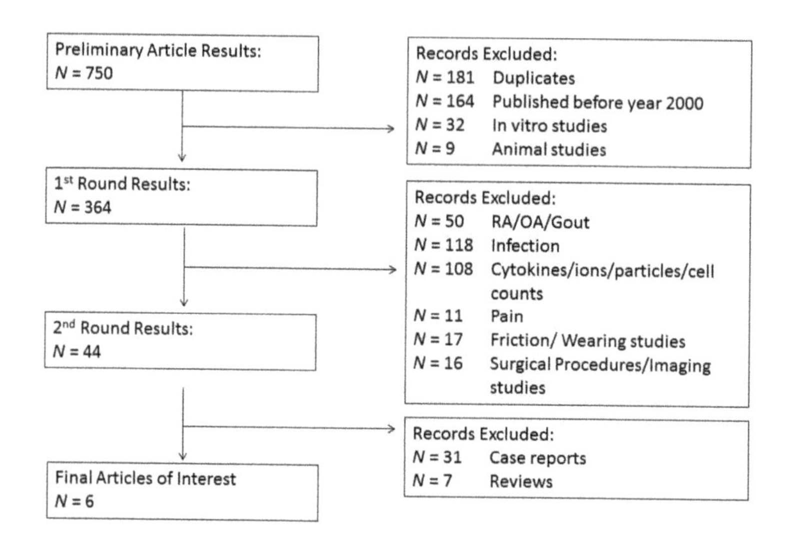

Figure 2. Flowchart of synovial lubrication article selection.

Six suitable papers were selected. Of these papers, four dealt specifically with the composition of the synovial fluid and the role this had on lubrication, while two discussed the role of increased temperature on lubricating protein degradation.

3. Results

3.1. *Synovial Lining Characterization*

Very few papers specifically described the histomorphology of the synovial intima in the periprosthetic capsular tissues and four of the five selected studies described the synovial-like membranes at interfaces. A common finding from those four studies of a synovial-like membrane at the bone-implant or bone-cement interface was the close resemblance to the true synovial lining membrane in cellular structure and cell type within the surface layer of excised implanted tissue. A range of histological presentations was also noted in each study, which was more pronounced in the implanted tissues compared with non-implanted tissues.

The study by Goldring *et al.* [17] was the first which characterized the new membrane that formed at the bone-cement interface into three distinct histological zones: (1) a synovial-like layer of lining cells at the cement surface; (2) sheets of histiocytes and giant cells in the subintima; and (3) a fibrous layer that blended into the bone and becomes continuous with the adjacent bone marrow. The membrane was lined with large polygonal cells, often with nuclei pointing away from the surface and a coating of fibrin around the surface. Cytoplasm was abundant and the nuclei were oval dark and uniform. Occasionally, multinucleated giant cells appeared within the lining cells. In a later study of prosthetic loosening after total hip arthroplasty, Goldring *et al.* confirmed the presence of a synovial-like lining with the same morphology in lining tissues adjacent to the implant [16]. Tissue cultures from implanted specimens showed three distinct cell types identified as stellate or dendritic cells, macrophage-like cells with phagocytosed latex or India-ink particles, and elongated fibroblast-like cells.

In a comparison of tissues from cemented, press-fit and biologic ingrowth prostheses revised due to aseptic loosening, Lennox *et al.* [19] described variations in the histomorphology and described three types of lining tissues. Type 1 linings comprised large, rounded or teardrop-shaped cells that were 1–3 cell layers thick. Type 2 membranes were composed of weakly stained collagenous matrix containing large dendritic-shaped cells with eccentrically located nuclei and long cellular processes extending toward the surface. Type 3 surfaces consisted of thick collagen fibers oriented perpendicular to the surface. All three surfaces were present on interface membranes from cemented, press-fit, and biologic ingrowth prostheses.

Lalor and Revell [18] compared the interface membrane of implanted tissue excised during revision surgery to normal synovium and rheumatoid arthritis tissue controls. They found that the new interface tissue had a layer of cells adjacent to the implant similar in appearance to normal synovial intima. This surface layer varied in thickness from 1 to 10 cells deep in a palisading structure, depending on each case and sample. Antibody staining for macrophage-like type A cells and fibroblast-like B cells was positive in the implanted tissue and showed the intermixing of the two cell types within the lining, similar to unimplanted tissue. Unlike the normal synovial lining and rheumatoid arthritis control tissues, separate distinct layers of macrophages and deeper fibroblasts were not seen in the implanted tissues, which could be because of the abundant macrophage infiltration of the subintimal area in response to wear debris.

In another comparison study of non-implanted and implanted tissues by Burkandt *et al.* [15] immunohistochemical staining was used to compare tissues from patients with low and high grade synovitis (inflammation), metal or polyethylene (PE) wear particle-induced synovitis and proliferative desquamative synovitis (associated with suspected metal hypersensitivity). The authors reasoned that, since proliferation of the synovial lining cell layer is a characteristic feature of autoimmune joint diseases, morphological changes of the synovial lining in periprosthetic tissues may indicate the presence of an immune response. The intimal lining of the PE wear particle-induced synovitis group was 1–2 cells thick, comparable to the unimplanted low grade synovitis group. By contrast the tissue from the high grade unimplanted synovitis tissue lining was five or more cells thick while the tissues from two metal-on-metal cases with extensive metal wear particle staining had a denuded cellular lining. The tissues from five cases with suspected metal hypersensitivity had a proliferating, viable synovial lining containing abundant fibroblast-like cells. Interestingly, the subsynovial layer of those tissues contained lymphocytes, an immune cell that was only seen in the high-grade rheumatoid synovitis tissues but not in the PE hip tissues. The authors suggested that the morphological similarities between tissues from patients revised for suspected metal hypersensitivity and patients with rheumatoid arthritis adds to the evidence that metal hypersensitivity shares characteristic morphological features with autoimmune diseases of the joints.

Figure 3 shows light micrographs of synovial tissue from an osteoarthritic joint prior to metal-on-polyethylene joint replacement (Figure 3A) and then following revision of that joint replacement (Figure 3B) in the same patient. Microscopic examination of both pre- and post-revision specimens in this patient shows an intimal/synovial lining with an underlying, non-specific, lymphoplasmacytic, chronic inflammatory infiltrate adjacent to, or centered around blood vessels. The post-revision specimen displays a denser collagenous subintimal layer (Figure 3B). Also present in the post-revision tissues are fragments of polarizable polyethylene with accompanying chronic inflammation including foreign body giant cells (Figure 3C).

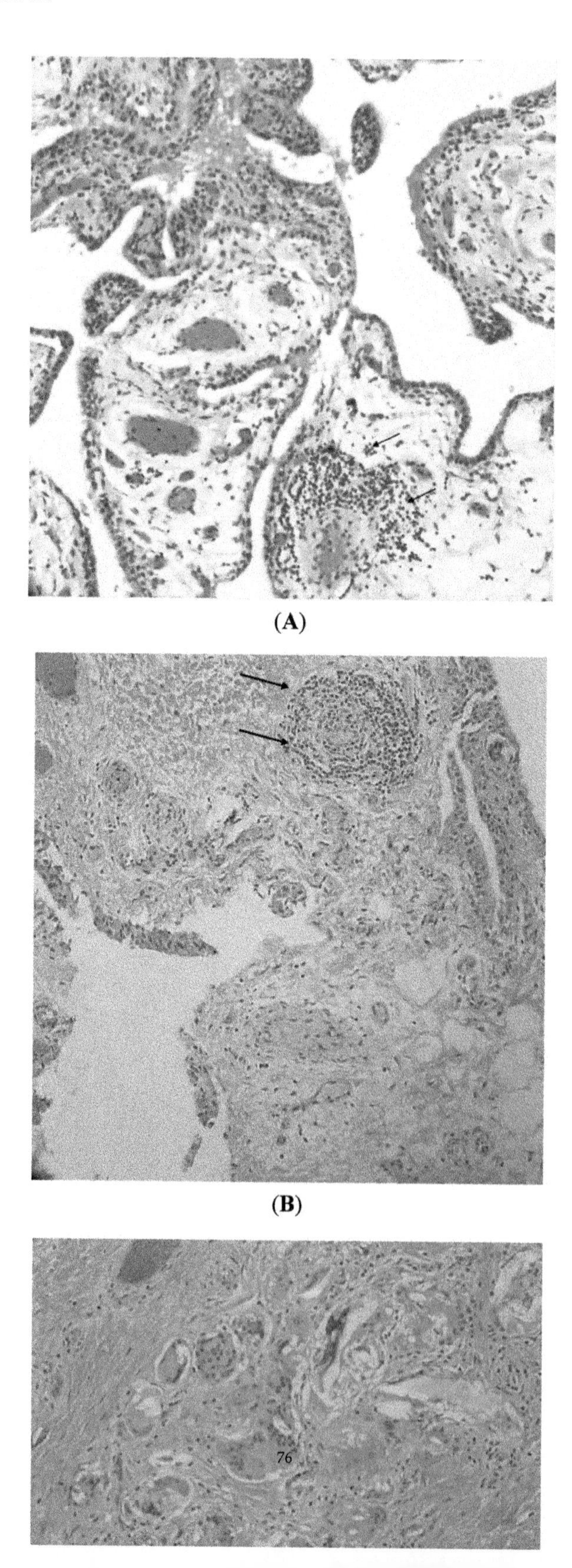

(**A**)

(**B**)

3.2. Synovial Fluid Properties

There is a large amount of information on the topic of synovial fluid properties and there is also a substantial body of literature on the tribology of artificial prosthetic joints. One review and analysis of the last decade of literature reiterated the complex nature of SF as a lubricant in both natural and artificial joints [42]. Two major mechanisms are theorized to play a role in both natural and artificial joints; boundary lubrication and fluid film lubrication. Macromolecules within the synovial fluid play a key role in both forms of lubrication. Another important mechanism of lubrication not addressed by the articles presented in this review is the theory of self-pressurized hydrostatic lubrication or biphasic lubrication. Cartilage has a biphasic nature in which a porous elastic matrix is infiltrated by interstitial fluid [43]. As the interstitial fluid is forced to flow through the permeable matrix a pressure gradient develops. The cartilage interstitial water supports most of the joint contact load. Pressurization of this fluid shifts most of the contact load away from the collagen-proteoglycan matrix resulting in a stress shielding effect and a low friction coefficient [43,44]. It should also be noted that with articulating cartilage and soft tissue, fluid pressurization is another distinct mechanism, besides boundary and fluid film lubrication, which is important. However, this review showed that there is little information about the characteristics of the fluid lubricating prosthetic joints or how alterations in implanted synovial tissue affect lubricant production, composition or efficiency.

In a study by Costa *et al.* [36] the constituents of synovial fluids that had been absorbed into the surfaces of ten revised PE hip implants (5 ethylene oxide/5 gamma ray sterilized in air) that were implanted for degenerative arthritis and replaced because of aseptic failure after 6–23 years were characterized. These products were extracted with cyclohexane then examined by mass spectrophotometry and Fourier transform infrared (FTIR) analysis was performed on the cross sections of the implants. Methyl esters of hexadecanoic and octadecanoic acids, of squalene, and of cholesterol were found in the extracts and a protein-like material was found on the surface of the implants. The diffusion of these molecules was postulated to cause the semi-crystalline polymer to become plasticized, especially at the surface where the greatest amount of absorbed material was found. This could cause changes to the ultimate tensile strength and ultimate elongation of the polymer, which could lead to a softer surface and decreased resistance to abrasion. The authors noted that accelerated simulated artificial joint testing may not allow time for the adsorption of synovial fluid constituents.

Joint fluid flow properties likely play an important role in fluid-film lubrication. A study by Mazzucco *et al.* [37] examined the flow properties of synovial fluid at the time of total knee arthroplasty (TKA) and at revision. Fifty-eight samples were obtained from patients undergoing index TKA surgery for osteoarthritis, 19 from revision TKA in other patients and two samples from aspirated effused joints that had undergone previous TKA. The volume of 22 samples from TKA, and five from revision TKA were insufficient for testing. Of the 14 samples from revision TKA, seven were revised because of wear and seven for mechanical problems not specific to wear. The average age of the tested patients was 70 years old (42 to 89 years). Synovial fluid at revision had lower viscosity in comparison to fluid at index TKA, which was more often within the normal range. However, the difference was not found to be statistically significant by the authors. There was variation in terms of viscosity in both pre- and post-revision groups and viscosity in both was decreased compared to synovial fluid from healthy young individuals. The authors suggested that, since hyaluronic acid (HA) is considered the major component of SF that affects its viscosity, the addition of HA to the lubricants used in joint simulation studies could provide an artificial joint fluid with viscosity more similar to natural SF. Previous studies performed on animals [45] have shown decreased HA concentrations after joint replacement which may explain the observations in these revision fluids.

Having explored fluid-film lubrication previously, Mazzucco *et al.* [38] explored the correlation of certain molecules associated with boundary lubrication and their association with flow properties. These boundary lubricants included phospholipids and proteins that passively diffuse through the joint capsule to reach the synovial fluid via filtration, and lubricin and superficial zone proteins (SZP) that are actively secreted into the joint fluid by synoviocytes. One hundred joint fluid samples were obtained, 77 from patients undergoing TKA for osteoarthritis (except 1 from post-traumatic arthritis), 20 from revision TKA, and 3 samples from aspirated effused joints that had previously undergone TKA. However, the majority of the samples (69) were insufficient in volume for analysis. Of the 14 revision samples examined, 10 were revised for wear related osteolysis and 4 for non-wear specific mechanical problems. A positive correlation was found between protein concentration and phospholipid concentration in pre and post-TKA synovial fluid. This indicated that the joint capsule functioned the same as before surgery and allowed diffusion of these macromolecules from interstitial fluid to the joint SF. A negative correlation was seen between these two components of SF and hyaluronic acid *i.e.*, when HA levels were low, protein and phospholipids were high, or when HA levels were high, protein and phospholipids were low. The authors offered two explanations for this. One was that when HA was low, the synovial membrane could compensate by allowing increased entry of proteins into the synovial fluid. Another explanation was that when protein content was high synoviocytes down-regulate production of HA. Seemingly the ability of the joint capsule to synthesize SF components and regulate macromolecule filtration is linked to one another. These findings were found in both pre and post-TKA specimens.

The surface-active phospholipids (SAPLs) in lubricin are believed to play an important role in boundary lubrication and analyzing the surfaces of revised implants for their presence was performed by Gale *et al.* [39]. The bearing surfaces of 38 revised total hips and 2 total knees, all with metal-on-polyethylene bearings were rinsed and analyzed by HPLC for phospholipids. In total eight different species of phosphatidylcholines were identified by the study. Palmitoyl linoleoyl phosphatidylcholine palmitoyl oleoyl phosphatidylcholine and stearoyl linoleoyl phosphatidylcholine were identified as the phospholipids most likely responsible for boundary lubrication. The authors recommended that a combination of saturated and unsaturated phosphatidylcholines be added to joint simulator lubricants to produce an artificial joint fluid that more closely simulated the boundary lubrication properties of normal SF.

In a novel study using an instrumented total hip prosthesis with a metal-on-polyethylene bearing, Bergmann *et al.* [40] measured the temperature resulting from friction at the bearing surfaces caused by walking and other physical activities. In part 1 of the study, temperatures inside two types of telemeterized hip prostheses were recorded at 9 locations along the prosthesis length. The implant comprised a titanium shaft, aluminum oxide ceramic head and a polyethylene cup. This non-cemented prosthesis was used in 4 patients, one of which has a second contralateral instrumented implant with a ceramic cup. Steady state temperatures were reached after 60 min of walking. The temperatures often rose above the critical level (42 °C) needed to cause synovial fluid protein degradation and precipitation. The joint fluid components lost to precipitation could result in decreased lubrication properties.

The authors noted that the volume of the synovial fluid and its lubricating function play a large role in the generation of heat in the active joint. If the acetabular cup material has good conductivity such as ceramic and metal, it would facilitate transfer of heat to the acetabular bone and away from the synovial fluid, capsule, and stem. The same is true of the stem, where a good heat conducting material like cobalt chromium alloy transfers heat away from the femoral head to the colder part of the implant. The authors suggested that an implant with better head and cup separation during the swing phase of walking would allow for better lubrication and heat dissipation. Bergmann *et al.* [41] noted that if hip simulators use a constant controlled temperature of 37 °C, such conditions may not mimic the rising and falling temperatures related to variable levels of *in vivo* joint activity.

4. Discussion

The purpose of this review of the literature was to examine (1) what is known about the synovial lining tissue histomorphology and (2) the composition of synovial fluid in joints following implantation of prosthetic implants.

Specifically, the primary focus of this review was to find papers in which the intimal layer of the synovial lining was characterized because type A and type B synoviocytes reside there, and they are the cells that contribute to the lubrication properties of the joint. Therefore, stringent criteria were set in our search for articles of interest. Many key implant retrieval articles described changes in the synovial tissue in general but were excluded because they focused on the subintimal layer and the inflammatory responses due to wear debris. Of note, several studies of the histological features of tissues from metal-on-metal total hips have reported the partial or complete loss of the synovial lining cells and their replacement by fibrin or by necrotic tissue [46–50]. How this might affect the lubrication of those bearings has not been addressed. In addition, many papers described the synovial lining in great detail but were excluded because the studies were focused on tissues excised from arthritic or diseased joints, not implanted joints [1–3,51].

The membranes that form at the interfaces of implants have been consistently described as resembling the normal synovial lining. With immunohistochemical methods, implanted synovial-like membrane cells tested positive for macrophage-like and fibroblast-like markers [15–19]. These findings imply that the formation of an interfacial membrane after implantation is a natural response to the influence of micromotion or from chemical mediators such as hyaluronic acid. The similarity in the intimal cell types and arrangement in membranes and joint linings has led to the conclusion that these tissues are likely to be capable of inflammatory cytokine production leading to bone loss and local tissue damage in the same way that the rheumatoid synovial pannus is responsible for local tissue destruction [17].

Under conditions of osteoarthritis and rheumatoid arthritis, synovitis is present and the increase in cell layer thickness, the excess of synoviocytes and increased fluid production are thought to contribute to the destruction of articular cartilage, and the formation of bone cysts. The thickening of fibrotic tissue in the synovial lining and dense cellular infiltrates of lymphocytes and monocytes are common morphological observations of the diseased synovium [52]. These changes in the synovial lining structure may transform the natural lubrication properties of the synovial fluid after joint replacement surgery. Delecrin *et al.* found that the level of total synovial fluid after total arthroplasty increased in a rabbit model, but the level of hyaluronic acid remained significantly lower than the internal controls [45] and the decrease in HA after implantation was found to be similar in humans as well [53].

For the selection of papers describing or characterizing implanted joint fluids, we excluded a large number of articles that focused on clinical management, synovial biomarkers, *in vitro* simulation studies and others that mentioned synovial fluid but not from the implanted joint. This left only a handful of studies of synovial fluids after arthroplasty. These studies reported that even though post-implantation SF properties are similar to pre-arthroplasty SF properties in some regards, significant alterations in composition and properties exist after arthroplasty surgery. Phospholipid overall contribution to boundary lubrications still remains unclear. Initial studies into its role utilized phospholipases, but it appears these early studies were contaminated with low levels of proteolytic enzymes. Later studies with proteolytic inhibitors did not show this increase in friction after phospholipase digestion. Studies performed *in vivo* in rabbits, however, have shown a reduced coefficient of frictions when DPPC liposomes and hyaluronic acid were delivered via intraarticular injections to damaged joints in comparison to only hyaluronic acid injections [54]. Studies have found that hyaluronic acid plays a key role in fluid film lubrication and the viscosity of the synovial fluid. Decreased production of HA by type B synoviocytes could explain the decreased viscosity in post arthroplasty fluid and a reciprocal increase of joint wear [37,38]. Lubricin has previously been identified as a key protein in boundary lubrication [55]. The role of species other than HA and lubricin, such as the various phosphatidylcholines, in post arthroplasty lubrication should be explored in future studies as they may have significance in boundary lubrication. These include hexadecanoic acid, octadecanoic acids, squalene, and cholesterol. [36] These proteins adsorb at the cobalt chromium surface to form thin, discontinuous deposited films *in vitro*. Because increasing protein content increases film thickness and can be directly correlated with femoral head wear, it seems likely that patient SF chemistry plays an important role in lubrication [56]. The choice of the optimal implant bearing materials for a given patient may, in future, be based on a better understanding of this individual SF chemistry.

One of the themes throughout the small number of studies characterizing post arthroplasty lubricants was the comparison with artificial joint simulator lubricants. Joint simulator studies have been performed for decades but there are often large differences between labs in the properties of the lubricants used to conduct the tests [42,57–59]. The conditions that these simulations are run under also differ markedly from joint function *in vivo* including the volume and temperature of the fluids. Several authors noted that artificial lubricant differs from post arthroplasty synovial fluid in many regards and suggested ways that this could be improved. For example, supplementation of hyaluronic acid, and a mixture of saturated and unsaturated phosphatidylcholine could lead to an artificial synovial fluid with properties more similarly seen *in vivo*. Brandt *et al.* suggest the addition of hyaluronic acid and phosphate-buffered saline to alpha-calf serum to be essential constituents in artificial lubricants [60]. Running the simulators with temperatures measured *in vivo* and for a longer duration could lead to tribological conditions that more closely resemble what an artificial joint experiences in the body [40].

In addition, higher serum degradation and larger wear particles were observed with smaller fluid volumes used for testing [61]. Despite the differences between *in vitro* and *in vivo* joint lubrication, the overall success of joint replacement components shows that the fundamental requirements for clinical use have been met.

5. Conclusions

There was a surprising lack of studies on implanted synovial tissue and the lubricating fluid it produces, although that tissue presumably plays an important role in the tribology and success of joint arthroplasty. The synovial lining tissue regenerates after implantation and produces a lubricating fluid that is sufficient in volume and lubricating constituents to allow the majority of joint replacements to function successfully, possibly for decades. The exact conditions for well-functioning implants are unclear; however, preserving the integrity of the joint including the synovial lining and natural lubricating fluid properties for joint arthroplasty may be key to successful implant survivorship. Whether the degree of implant wear or if some cases of failure can be attributed to any synovial tissue or lubricant deficiency are questions that remain to be answered and warrant future investigation.

Author Contributions: Michael Shang Kung and John Markantonis performed the systematic review. Patricia Campbell and Scott D. Nelson designed the study, assisted with the literature search and reviewed the histology. All authors were involved in the manuscript preparation and editing for final submission.

Conflicts of Interest: The authors declare no conflict of interest.

References

1. Ghadially, F.N. *Fine Structure of Synovial Joints: A Text and Atlas of the Ultrastructure of Normal and Pathological Articular Tissues*; Butterworths: London, UK; Boston, MA, USA, 1983.
2. Ghadially, F.N.; Roy, S. *Ultrastructure of Synovial Joints in Health and Disease*; Butterworths: London, UK, 1969.
3. Key, J.A. *The Synovial Membrane of Joints and Bursae*, 2nd ed.; Paul, B., Hober: New York, NY, USA, 1932.
4. Bronner, F.; Farach-Carson, M.C. *Bone and Osteoarthritis*; Springer-Verlag: London, UK, 2007; Volume 4.
5. Pavlovich, R.I.; Lubowitz, J. Current concepts in synovial tissue of the knee joint. *Orthopedics* **2008**, *31*, 160–163. [CrossRef] [PubMed]
6. Athanasou, N.A.; Quinn, J.; Heryet, A.; Puddle, B.; Woods, C.G.; McGee, J.O. The immunohistology of synovial lining cells in normal and inflamed synovium. *J. Pathol.* **1988**, *155*, 133–142. [CrossRef] [PubMed]
7. Barland, P.; Novikoff, A.B.; Hamerman, D. Electron microscopy of the human synovial membrane. *J. Cell Biol.* **1962**, *14*, 207–220. [CrossRef] [PubMed]
8. Iwanaga, T.; Shikichi, M.; Kitamura, H.; Yanase, H.; Nozawa-Inoue, K. Morphology and functional roles of synoviocytes in the joint. *Arch. Histol. Cytol.* **2000**, *63*, 17–31. [CrossRef] [PubMed]
9. Mapp, P.I.; Revell, P.A. Fibronectin production by synovial intimal cells. *Rheumatol. Int.* **1985**, *5*, 229–237. [CrossRef] [PubMed]

10. Matsubara, T.; Spycher, M.A.; Ruttner, J.R.; Fehr, K. The ultrastructural localization of fibronectin in the lining layer of rheumatoid arthritis synovium: The synthesis of fibronectin by type B lining cells. *Rheumatol. Int.* **1983**, *3*, 75–79. [CrossRef] [PubMed]
11. Yielding, K.L.; Tomkins, G.M.; Bunim, J.J. Synthesis of hyaluronic acid by human synovial tissue slices. *Science* **1957**, *125*, 1300. [CrossRef] [PubMed]
12. Hogg, N.; Palmer, D.G.; Revell, P.A. Mononuclear phagocytes of normal and rheumatoid synovial membrane identified by monoclonal antibodies. *Immunology* **1985**, *56*, 673–681. [PubMed]
13. Depalma, A.F. *Diseases of the Knee: Management in Medicine and Surgery*, 1st ed.; J.B. Lippincott Company: Philadelphia, PA, USA, 1954.
14. Key, J.A. The reformation of synovial membrane in the knees of rabbits after synovectomy. *J. Bone Joint Surg.* **1925**, *7*, 793–813.
15. Burkandt, A.; Katzer, A.; Thaler, K.; Von Baehr, V.; Friedrich, R.E.; Ruther, W.; Amling, M.; Zustin, J. Proliferation of the synovial lining cell layer in suggested metal hypersensitivity. *In Vivo* **2011**, *25*, 679–686. [PubMed]
16. Goldring, S.R.; Jasty, M.; Roelke, M.S.; Rourke, C.M.; Bringhurst, F.R.; Harris, W.H. Formation of a synovial-like membrane at the bone-cement interface. Its role in bone resorption and implant loosening after total hip replacement. *Arthritis Rheum.* **1986**, *29*, 836–842.
17. Goldring, S.R.; Schiller, A.L.; Roelke, M.; Rourke, C.M.; O'Neil, D.A.; Harris, W.H. The synovial-like membrane at the bone-cement interface in loose total hip replacements and its proposed role in bone lysis. *J. Bone Joint Surg.* **1983**, *65*, 575–584. [PubMed]
18. Lalor, P.A.; Revell, P.A. The presence of a synovial layer at the bone-implant interface: An immunohistological study demonstrating the close similarity to true synovium. *Clin. Mater.* **1993**, *14*, 91–100. [CrossRef]
19. Lennox, D.W.; Schofield, B.H.; McDonald, D.F.; Riley, L.H., Jr. A histologic comparison of aseptic loosening of cemented, press-fit, and biologic ingrowth prostheses. *Clin. Orthop. Relat. Res.* **1987**, 171–191.
20. Drachman, D.B.; Sokoloff, L. Role of movement in embryonic joint development. *Dev. Biol.* **1966**, *14*, 401–420. [CrossRef]
21. Engh, C.A.; Oconnor, D.; Jasty, M.; Mcgovern, T.F.; Bobyn, J.D.; Harris, W.H. Quantification of implant micromotion, strain shielding, and bone-resorption with porous-coated anatomic medullary locking femoral prostheses. *Clin. Orthop. Relat. Res.* **1992**, *285*, 13–29. [PubMed]
22. Konttinen, Y.T.; Li, T.F.; Mandelin, J.; Ainola, M.; Lassus, J.; Virtanen, I.; Santavirta, S.; Tammi, M.; Tammi, R. Hyaluronan synthases, hyaluronan, and its CD44 receptor in tissue around loosened total hip prostheses. *J. Pathol.* **2001**, *194*, 384–390. [CrossRef] [PubMed]
23. Edwards, J.C.; Sedgwick, A.D.; Willoughby, D.A. The formation of a structure with the features of synovial lining by subcutaneous injection of air: An *in vivo* tissue culture system. *J. Pathol.* **1981**, *134*, 147–156. [CrossRef] [PubMed]
24. Pap, G.; Machner, A.; Rinnert, T.; Horler, D.; Gay, R.E.; Schwarzberg, H.; Neumann, W.; Michel, B.A.; Gay, S.; Pap, T. Development and characteristics of a synovial-like interface membrane around cemented tibial hemiarthroplasties in a novel rat model of aseptic prosthesis loosening. *Arthritis Rheum.* **2001**, *44*, 956–963. [CrossRef] [PubMed]
25. Sedgwick, A.D.; Sin, Y.M.; Edwards, J.C.W.; Willoughby, D.A. Increased inflammatory reactivity in newly formed lining tissue. *J. Pathol.* **1983**, *141*, 483–495. [CrossRef] [PubMed]
26. Radin, E.L.; Paul, I.L.; Swann, D.A.; Schottstaedt, E.S. Lubrication of synovial membrane. *Ann. Rheum. Dis.* **1971**, *30*, 322–325. [CrossRef] [PubMed]
27. Konttinen, Y.T.; Zhao, D.; Beklen, A.; Ma, G.; Takagi, M.; Kivela-Rajamaki, M.; Ashammakhi, N.; Santavirta, S. The microenvironment around total hip replacement prostheses. *Clin. Orthop. Relat. Res.* **2005**, *430*, 28–38. [CrossRef] [PubMed]
28. Hills, B.A.; Butler, B.D. Surfactants identified in synovial fluid and their ability to act as boundary lubricants. *Ann. Rheum. Dis.* **1984**, *43*, 641–648. [CrossRef] [PubMed]
29. Seror, J.; Zhu, L.; Goldberg, R.; Day, A.J.; Klein, J. Supramolecular synergy in the boundary lubrication of synovial joints. *Nat. Commun.* **2015**, *6*. [CrossRef] [PubMed]
30. Radin, E.L.; Swann, D.A.; Weisser, P.A. Separation of a hyaluronate-free lubricating fraction from synovial fluid. *Nature* **1970**, *228*, 377–378. [CrossRef] [PubMed]
31. McCutchen, C.W. Joint lubrication. *Bull. Hosp. Jt. Dis. Orthop. Inst.* **1983**, *43*, 118–129.

32. Balazs, E.A. *The Physical Properties of Synovial Fluid and the Specific Role of Hyaluronic Acid*; J B Lippincott: Philadelphia, PA, USA, 1982.
33. Balazs, E.A.; Watson, D.; Duff, I.F.; Roseman, S. Hyaluronic acid in synovial fluid. I. Molecular parameters of hyaluronic acid in normal and arthritis human fluids. *Arthritis Rheum.* **1967**, *10*, 357–376.
34. Ogston, A.G.; Stanier, J.E. The physiological function of hyaluronic acid in synovial fluid; viscous, elastic and lubricant properties. *J. Physiol.* **1953**, *119*, 244–252. [CrossRef] [PubMed]
35. Wang, M.; Liu, C.; Thormann, E.; Dedinaite, A. Hyaluronan and phospholipid association in biolubrication. *Biomacromolecules* **2013**, *14*, 4198–4206. [CrossRef] [PubMed]
36. Costa, L.; Bracco, P.; del Prever, E.B.; Luda, M.P.; Trossarelli, L. Analysis of products diffused into UHMWPE prosthetic components *in vivo*. *Biomaterials* **2001**, *22*, 307–315. [CrossRef] [PubMed]
37. Mazzucco, D.; McKinley, G.; Scott, R.D.; Spector, M. Rheology of joint fluid in total knee arthroplasty patients. *J. Orthop. Res.* **2002**, *20*, 1157–1163. [CrossRef] [PubMed]
38. Mazzucco, D.; Scott, R.; Spector, M. Composition of joint fluid in patients undergoing total knee replacement and revision arthroplasty: Correlation with flow properties. *Biomaterials* **2004**, *25*, 4433–4445. [CrossRef] [PubMed]
39. Gale, L.R.; Chen, Y.; Hills, B.A.; Crawford, R. Boundary lubrication of joints: Characterization of surface-active phospholipids found on retrieved implants. *Acta Orthop.* **2007**, *78*, 309–314. [CrossRef] [PubMed]
40. Bergmann, G.; Graichen, F.; Rohlmann, A.; Verdonschot, N.; van Lenthe, G.H. Frictional heating of total hip implants. Part 1: Measurements in patients. *J. Biomech.* **2001**, *34*, 421–428.
41. Bergmann, G.; Graichen, F.; Rohlmann, A.; Verdonschot, N.; van Lenthe, G.H. Frictional heating of total hip implants. Part 2: Finite element study. *J. Biomech.* **2001**, *34*, 429–435.
42. Ghosh, S.; Choudhury, D.; Das, N.S.; Pingguan-Murphy, B. Tribological role of synovial fluid compositions on artificial joints—A systematic review of the last 10 years. *Lubr. Sci.* **2014**, *26*, 387–410. [CrossRef]
43. Bonnevie, E.D.; Baro, V.J.; Wang, L.; Burris, D.L. Fluid load support during localized indentation of cartilage with a spherical probe. *J. Biomech.* **2012**, *45*, 1036–1041. [CrossRef] [PubMed]
44. Caligaris, M.; Ateshian, G.A. Effects of sustained interstitial fluid pressurization under migrating contact area, and boundary lubrication by synovial fluid, on cartilage friction. *Osteoarthritis Cartilage* **2008**, *16*, 1220–1227. [CrossRef] [PubMed]
45. Delecrin, J.; Oka, M.; Takahashi, S.; Yamamuro, T.; Nakamura, T. Changes in joint fluid after total arthroplasty. A quantitative study on the rabbit knee joint. *Clin. Orthop. Relat. Res.* **1994**, *307*, 240–249.
46. Campbell, P.; Ebramzadeh, E.; Nelson, S.; Takamura, K.; De Smet, K.; Amstutz, H.C. Histological features of pseudotumor-like tissues from metal-on-metal hips. *Clin. Orthop. Relat. Res.* **2010**, *468*, 2321–2327. [CrossRef] [PubMed]
47. Davies, A.P.; Willert, H.G.; Campbell, P.A.; Learmonth, I.D.; Case, C.P. An unusual lymphocytic perivascular infiltration in tissues around contemporary metal-on-metal joint replacements. *J. Bone Joint Surg.* **2005**, *87*, 18–27. [CrossRef] [PubMed]
48. Grammatopoulos, G.; Pandit, H.; Kamali, A.; Maggiani, F.; Glyn-Jones, S.; Gill, H.S.; Murray, D.W.; Athanasou, N. The correlation of wear with histological features after failed hip resurfacing arthroplasty. *J. Bone Joint Surg.* **2013**, *95*, e81. [CrossRef] [PubMed]
49. Howie, D.W.; Cain, C.M.; Cornish, B.L. Pseudo-abscess of the psoas bursa in failed double-cup arthroplasty of the hip. *J. Bone Joint Surg.* **1991**, *73*, 29–32.
50. Willert, H.G.; Buchhorn, G.H.; Fayyazi, A.; Flury, R.; Windler, M.; Koster, G.; Lohmann, C.H. Metal-on-metal bearings and hypersensitivity in patients with artificial hip joints. A clinical and histomorphological study. *J. Bone Joint Surg.* **2005**, *87*, 28–36. [CrossRef]
51. Hirohata, K.; Kobayashi, I. Fine structures of the synovial tissues in rheumatoid arthritis. *Kobe J. Med. Sci.* **1964**, *10*, 195–225. [PubMed]
52. Haraoui, B.; Pelletier, J.P.; Cloutier, J.M.; Faure, M.P.; Martel-Pelletier, J. Synovial membrane histology and immunopathology in rheumatoid arthritis and osteoarthritis. *In vivo* effects of antirheumatic drugs. *Arth. Rheum.* **1991**, *34*, 153–163.
53. Walker, P.S. A comparison of normal and artificial human joints. *Acta Orthop. Belg.* **1973**, *39*, 43–54. [PubMed]
54. Jay, G.D.; Waller, K.A. The biology of lubricin: Near frictionless joint motion. *Matrix Boil. J. Int. Soc. Matrix Biol.* **2014**, *39*, 17–24. [CrossRef]

55. Swann, D.A.; Silver, F.H.; Slayter, H.S.; Stafford, W.; Shore, E. The molecular structure and lubricating activity of lubricin isolated from bovine and human synovial fluids. *Biochem. J.* **1985**, *225*, 195–201. [PubMed]
56. Fan, J.; Myant, C.; Underwood, R.; Cann, P. Synovial fluid lubrication of artificial joints: Protein film formation and composition. *Faraday Discuss.* **2012**, *156*, 69–85. [CrossRef] [PubMed]
57. Gispert, M.P.; Serro, A.P.; Colaco, R.; Saramago, B. Friction and wear mechanisms in hip prosthesis: Comparison of joint materials behaviour in several lubricants. *Wear* **2006**, *260*, 149–158. [CrossRef]
58. Roba, M.; Bruhin, C.; Ebneter, U.; Ehrbar, R.; Crockett, R.; Spencer, N.D. Latex on glass: An appropriate model for cartilage-lubrication studies? *Tribol. Lett.* **2010**, *38*, 267–273. [CrossRef]
59. Wang, A.; Essner, A.; Schmidig, G. The effects of lubricant composition on *in vitro* wear testing of polymeric acetabular components. *J. Biomed. Mater. Res. B Appl. Biomater.* **2004**, *68*, 45–52. [CrossRef] [PubMed]
60. Brandt, J.M.; Briere, L.K.; Marr, J.; MacDonald, S.J.; Bourne, R.B.; Medley, J.B. Biochemical comparisons of osteoarthritic human synovial fluid with calf sera used in knee simulator wear testing. *J. Biomed. Mater. Res. A* **2010**, *94*, 961–971. [PubMed]
61. Reinders, J.; Sonntag, R.; Kretzer, J.P. Synovial fluid replication in knee wear testing: An investigation of the fluid volume. *J. Orthop. Res.* **2015**, *33*, 92–97. [CrossRef] [PubMed]

Review

Wear Performance of UHMWPE and Reinforced UHMWPE Composites in Arthroplasty Applications: A Review

Juan C. Baena *, Jingping Wu and Zhongxiao Peng

School of Mechanical and Manufacturing Engineering, The University of New South Wales, Sydney NSW 2052, Australia; j.p.wu@unsw.edu.au (J.W.); z.peng@unsw.edu.au (Z.P.)

* Author to whom correspondence should be addressed; juan.baenavargas@student.unsw.edu.au; Tel.: +612-9358-4249.

Academic Editors: Amir Kamali and J. Philippe Kretzer
Received: 27 February 2015; Accepted: 4 May 2015; Published: 18 May 2015

Abstract: As the gold standard material for artificial joints, ultra-high-molecular-weight polyethylene (UHMWPE) generates wear debris when the material is used in arthroplasty applications. Due to the adverse reactions of UHMWPE wear debris with surrounding tissues, the life time of UHMWPE joints is often limited to 15–20 years. To improve the wear resistance and performance of the material, various attempts have been made in the past decades. This paper reviews existing improvements made to enhance its mechanical properties and wear resistance. They include using gamma irradiation to promote the cross-linked structure and to improve the wear resistance, blending vitamin E to protect the UHMWPE, filler incorporation to improve the mechanical and wear performance, and surface texturing to improve the lubrication condition and to reduce wear. Limitations of existing work and future studies are also identified.

Keywords: UHMWPE; wear resistance; mechanical properties; gamma irradiation; fillers; surface engineering

---

1. Introduction

Artificial joints are designed for a partial or complete replacement of articular joints affected by a degenerative disease known as osteoarthritis (OA). The estimated life span of a conventional artificial knee joint (AKJ) is around 15 to 20 years [1], mainly determined by the mechanical properties and wear resistance of the material used and the specific conditions of each patient [2]. Due to a continuous increase in orthopaedic surgical procedures and the fact that more young people are suffering OA [3], extending the lifespan of artificial joints is imperative and has become the most challenging issue in the improvement of artificial joint performance.

The degradation of artificial joints (AJ) is caused by many factors. Since AJ operate under sliding and rolling conditions [4], the surface degradation process is realised through specific wear mechanisms such as abrasion, adhesion and fatigue [5]. An increase in the external load and/or the surface roughness of the contact material(s) intensifies the wear process. The mechanical and physical properties of the materials also play an important role in the degradation process. The interaction of all these variables defines the tribological system of AJ and its degradation process.

The lifespan of AJ is closely linked to the wear resistance of its components. The wear of AJ affects the performance of the joint by causing hazardous biologic reactions in the body due to micron and sub-micron wear particles generated in the process [6]. Ultra-high-molecular-weight polyethylene (UHMWPE) has been considered as the gold standard material used in arthroplasty applications [7], especially for knee and hip artificial implant procedures. The UHMWPE, a component of polymer on metal (POM) joints, is a material subjected to severe material degradation and damage under sliding conditions due to its relatively high ductility, and relatively low hardness, Young's module and stiffness [8]. Large UHMWPE particles are not easily ingested by cells, proteins and/or body fluids [9], and, consequently, are less harmful than small sized wear debris. Small particles have a negative impact on the biological body system, affecting the integrity of cells, proteins and/or body fluids [9].

To enhance the performance of UHMWPE in terms of reducing its wear rate and wear particle generation, attempts have been made to improve the life span of the component. For example, the Hymaler, a high crystalline UHMWPE (with 73.2%) was identified as a potential material for arthroplasty application, since this material presents a high resistance to fatigue and creep propagation. However, it is susceptible to oxidation related degradation, which affects its clinical performance. This material was replaced by Maraton crosslinked UHMWPE [10] in 1997. Regarding to UHMWPE composite, it also has been considered as a potential alternative to improve the wear performance of AJ. The UHMWPE reinforced with carbon fibers (CFR-UHMWPE), named Poly II, was used in orthopaedic implants in the 1970s. This composite was discontinued due to evidences on reduced crack resistance, rupture of the fibers on the surface and other issues [11].

Existing works are based on the following findings. It is well recognised that the mechanical properties of UHMWPE influence its load-bearing capacity [10]. Moreover, the wear resistance of an articular cartilage (AC) is related to the efficiency of its lubrication system [11]. Even though UHMWPE is considered to have a low friction during a sliding process, its low affinity between the synovial fluid and the polyethylene surface limits the effectiveness of lubrication of the component [12]. That low affinity is related to surface wettability which is described by the inclination of the fluid to cover the surface. A material with a good wettability tends to form a tribofilm on the surface that has the capability of carrying load and avoiding surface contact. An enhancement in the wettability of UHMWPE can improve the lubrication conditions, which would be reflected on the wear resistance performance [13]. The surface topographies of the material play an important role in the lubrication system and affect the wear rate [14]. Furthermore, studying the histology and surface appearance of degraded UHMWPE components are important to understand the most relevant wear mechanisms that affect the component integrity.

To increase the life span of UHMWPE AJs, an improvement in the mechanical properties and/or surface and lubrication conditions can be achieved by various means including modifying the polymer structure, embedding filling materials to strengthen certain mechanical properties, and fabricating designed surface patterns on the outermost surface to improve lubrication conditions. This paper reviews reported studies on the improvement of the wear performance of UHMWPE composites for arthroplasty applications.

2. Mechanical Properties and Manufacturing Process of UHMWPE

UHMWPE is a polymer with outstanding mechanical and physical properties, including good wear and impact resistance, chemical inertness, lubricity and biocompatibility [15]. The mechanical and physical properties are intrinsic properties of the material, which can be modified in a manufacturing process. UHMWPE is widely used in industrial and clinical applications. Nowadays, around 2 million UHMWPE artificial joints are implanted every year [16]. The requirements of UHMWPE used for arthroplasty application are specified in ASTM standard F648 and ISO standard 5834-1. A raw material of this specific polyethylene is produced by Ticona-Celanese in two different powder classifications GUR1020 and GUR1050. They are classified according to their molecular weights, being around 3.5×10^6 g/mol and $5.5–6.0 \times 10^6$ g/mol for GUR1020 and GUR1050, respectively [17].

Compression moulding and ram extrusion are the two widely performed manufacturing processes to form bulk UHMWPE pieces [18]. The compression moulding was the first manufacturing process used on UHMWPE powder in the 1950s and this manufacturing method is still being used by two companies, Orthoplastics and MediTECH, who produce compression mould sheets of GUR 1020 and 1050. In contrast, the ram extrusion of UHMWPE has been used for over 30 years and this process is used by a few companies to turn GUR 1020 and 1050 into ram extruded UHMWPE. Using this process, UHMWPE rods can be manufactured in a wide range of diameters, being the most commercial products used for orthopaedic purpose from 20 to 80 mm in diameter [19]. Due to its high molecular weight, the UHMWPE has a high viscosity even above the melt point, making this material complex to be processed by conventional screw extrusion [20]. After a UHMWPE block is formed, it is machined to an orthopaedic component.

Since the wear resistance is closely correlated to the mechanical properties, the UHMWPE mechanical properties have been modified using different methods, including the employed manufacturing processes. Kurtz [21] reported that the tensile yield and ultimate tensile properties did not present significant differences between the polyethylene types and manufacturing processes, while the elongation to failure was significantly different between the two processes and/or materials. The differences in the mechanical properties are presented in Figure 1.

3. Improvements on the Wear Resistance of UHMWPE

Studies on the modification of UHMWPE have been performed in order to improve its wear resistance and consequently its clinical performance. This section reviews a number of existing modification techniques including crystallinity, cross-linking, adding antioxidants and reinforced filling materials for enhancing mechanical properties, and surface engineering for improving its wear resistance.

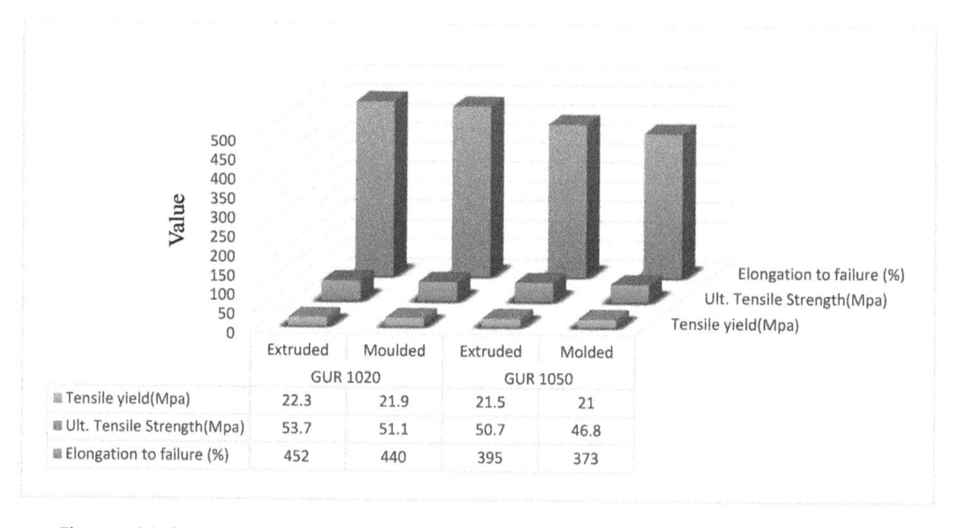

	GUR 1020 Extruded	GUR 1020 Moulded	GUR 1050 Extruded	GUR 1050 Molded
Tensile yield(Mpa)	22.3	21.9	21.5	21
Ult. Tensile Strength(Mpa)	53.7	51.1	50.7	46.8
Elongation to failure (%)	452	440	395	373

Figure 1. Mechanical properties of UHMWPE extruded and moulded. The data were taken from [21].

3.1. *Crystallinity of UHMWPE*

The mechanical and physical properties of UHMWPE are influenced by its crystallinity, which is related to the atomic or molecular arrangement, conforming a specific crystal structure [22]. The crystallinity of polymers is attributed to the specific arrangement of the molecular chains, forming crystalline regions or crystalline lamellae which are chain-folded molecules. Polymers are not totally crystalline and these lamellae are normally surrounded by random distributed molecules with amorphous regions [23].

Since the mechanical and physical properties of UHMWPE are highly affected by its microstructure and grade of crystallinity, it is expected that the wear performance can be improved by altering the microstructural conditions. Karuppiah *et al.* [1] evaluated the influence of crystallinity on the wear resistance of UHMWPE and reported that, with increasing the crystallinity, the friction force and the scratch depth tended to decrease. Their study suggested that the wear resistance increased with the degree of crystallinity. Similar results were reported by Wang and Ge [24] through varying mould pressures during the sample formation process. Furthermore, it was reported that the degree of crystallinity can be increased by controlling the material formation pressure and temperature. Heating the UHMWPE (with a molecular weight over 400,000) under high pressure followed by a posterior cooling process induces the formation of the crystalline structure. Lamellae with a thickness less than 12 nm were formed [25]. It was also reported that by holding the temperature above the melting point (around 110 °C) the chains of the polyethylene are able to fold and the crystalline lamellae is formed, resulting in an increase in the crystallinity of the material [1].

The degree of crystallinity in the material is estimated using differential scanning calorimetry (DSC) which determines the temperatures when the material experiences phase transitions such as crystallization and melting point. The phase transitions are evidenced by peaks at a specific temperature as shown in Figure 2 in [26]. The UHMWPE melting temperature is revealed by a sudden increase of the heat at around 137 °C. The degree of crystallinity (ΔXc) of UHMWPE is calculated based on the melting heat of crystallinity, ΔHm, which is the energy per unit mass required to melt the lamellae in the material [26]. The crystalline regions can be detected using transmission electron microscopy (TEM), which identifies the lamellae based on the white and gray contrast made with uranyl acetate. In TEM images the lamellae are revealed as white lines with defined dark borders and the amorphous region with grey contrast [16].

As stated before, the mechanical properties and the wear performances of UHMWPE are affected by its microstructure. Being a semicrystalline material, the degree of crystallinity of UHMWPE was estimated between 39% and 75% [17]. The lack of full crystallization imposes a limitation on its mechanical property enhancement. It is also known that the crystallinity of polymers is affected by temperature. It was found that, above 60 to 90 °C, small crystallites could be dissolved [16], which might affect the mechanical performance at relatively low temperatures.

3.2. Cross-Linked UHMWPE Using Gamma Irradiation

Radiation crosslinking in UHMWPE was developed using high-dose radiation to reduce the ductility of the polymer, to increase the hardness and to improve the wear resistance of UHMWPE [27,28]. Gamma radiation and electron beam (e-beam) radiation are commonly used as an ionizing radiation source to generate free radicals on the polymer. The recombination of some free radicals form the crosslinks, which reduce the chain stretch of the material, and consequently, the plastic deformation is significantly reduced. The wear resistance is increased with the radiation dose, as can be seen in Figure 2. However, a high radiation dose accelerates the oxidation, which increases the failure of the material [29]. This is because excessive free radicals migrate to the interface (crystalline/amorphous) and react with diffused oxygen, causing oxidation and brittleness. The oxidation increases the wear rate, and consequently, the rate of particulate wear of the orthopaedic component [20]. Green *et al.* [30] stated that even though crosslinking can reduce the wear volume, particles released from an irradiated UHMWPE can be more biologically active and smaller than the particles from a non-irradiated one. These small particles increase the probability of osteolysis.

There are different alternatives to quench the free radicals and avoid oxidation. One approach is a thermal treatment after irradiation. It is stated that heating the material below the melt point can preserve the crystallinity and mechanical properties, but it does not completely eliminate the excess of free radicals [25]. In contrast, by heating the polymer to above the melting point, the crystalline regions are eliminated and the free radicals present good mobility. This combined effect improves the oxidative stability of the irradiated UHMWPE but reducing its crystallinity [31].

Electron beam (e-beam) and gamma irradiation produce similar reactions in the polymeric chains. In comparison to gamma irradiation, the penetration of the e-beam radiation is limited. However, since the e-beam irradiation process is considered adiabatic, it allows a major control of the temperature increase during the irradiation [31]. Both techniques are used for the irradiation treatment of commercially available UHMWPE for clinical use. Table 1 shows the irradiation processes of some trademarks and their corresponding mechanical properties.

Even when irradiated UHMWPE presents a significant improvement on the wear resistance, the susceptibility to oxidation due to free radicals is a concern. The thermal treatment after the irradiation process can increase the oxidation stability of UHMWPE but oxidation is not completely controlled. An additional and widely used alternative to reduce the free radicals after irradiation is blending Vitamin-E with the UHMWPE, which will be reviewed in the next section.

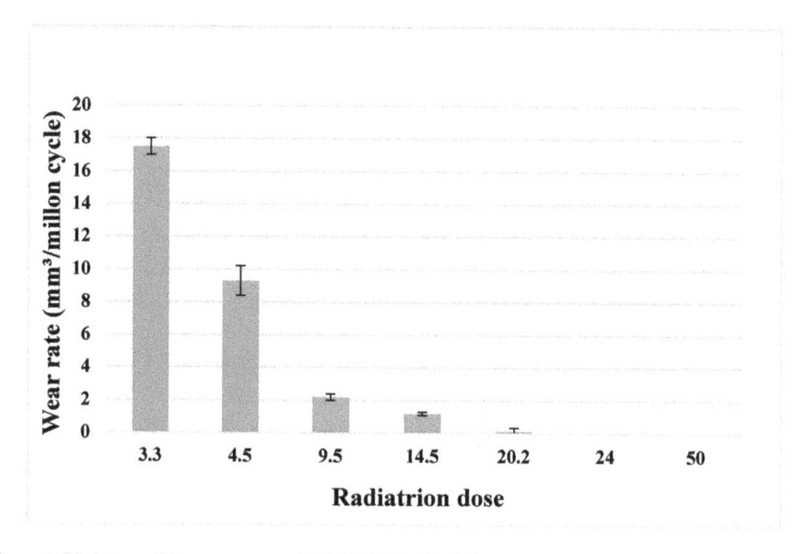

Figure 2. Variation of the wear rates of UHMWPE with different gamma radiation doses (1Mrad = 10 kGy). The data were taken from McKellop *et al.* [32].

3.3. *Vitamin-E-Stabilised UHMWPE*

As stated in the previous section, the crosslinking process induces the formation of free radicals in the crystalline phase, making the material susceptible to oxidation. These residual free radicals can migrate to the crystalline/amorphous interface and cause oxidation and reduction of UHMWPE elongation at break, tensile strength, fracture toughness and fatigue crack propagation resistance [33–38]. Therefore, to improve the oxidation resistance and the mechanical and fatigue strengths of cross-linked UHMWPE, antioxidant vitamin E is used to stabilise the radiation-induced free radicals [39–42]. The main role of vitamin E is to react with free radicals and protect the UHMWPE [39,43].

Table 1. Irradiation techniques employed on commercially available UHMWPE for clinical use and comparison between mechanical properties and the irradiation processes using GUR 1050 [31].

Commercial Names	Irradiation Technique	Irradiation Temp	Total Radiation Dose (kGy)	Post-Irradiation Thermal Treatment	Yield Strength (Mpa) (±SD)	Ultimate Tensile Strength (Mpa) (±SD)	True Elongation to Break (%)
UHMWPE unirradiated	-	-	-	-	21	50	340
Longevity [a] (Zimmer) [b]	E-beam	40 °C	100	Melted	21 ± 1.0	43 ± 9.8	240 ± 35
Durasul (Zimmer)		125 °C	95	Melted	20 ± 0.7	30 ± 7.1	280 ± 74
Crossfire (Stryker Howmedica)	Gamma	RT	105	130 °C	24 ± 1.3	48 ± 7.2	280 ± 37
Marathon (DePuy/J and J)		RT	50	Melted	21 ± 1.5	56 ± 5.7	290 ± 14
XLPE (Smith & Nephew)		RT	100	Melted	20 ± 1.3	56 ± 7.1	300 ± 20

RT = Room Temperature; a = trademark; b = Manufacturer.

3.3.1. Methods of Vitamin E Blending

Vitamin E can be incorporated into UHMWPE using two methods. The first method is diffusing vitamin E into UHMWPE after radiation crosslinking [44,45]. This method incorporates vitamin E into UHMWPE using a two-step process. The first step is post-radiation doping with vitamin E, and the second step is a further homogenization into inert atmosphere [44,45]. The homogenization step is necessary since it can achieve an adequate antioxidant concentration throughout the implants. Using this method, the crosslinking efficiency of UHMWPE is not limited because vitamin E is not incorporated during irradiation. Furthermore, the amount of incorporated vitamin E is not affected by the crosslink density. Due to the saturation limit of the vitamin E blended UHMWPE at body temperature, the vitamin E concentration is approximately 0.7 wt% [46]. However, the UHMWPE has a risk of oxidation during irradiation and in storage until vitamin E is incorporated. The vitamin E-stabilised cross-linked UHMWPE, which is produced using this method, has been clinically used since 2007 [44,45,47–49].

The second method is blending a liquid antioxidant into the UHMWPE resin powder before the mixture consolidates into a near-implant form and radiation crosslinking is performed [39,50–52]. This method is easier and takes shorter time to obtain uniform vitamin E concentration throughout the implant compared with the first method [53]. However, the presence of vitamin E during the irradiation process reduces the efficiency of UHMWPE crosslinking, because vitamin E itself can act as a free radical scavenger [52]. Therefore, the concentration of vitamin E and the radiation dose must be optimized to achieve a good wear and oxidation resistance. The vitamin E concentration in the blend is limited to less than 0.3 wt% [54], and the radiation dose of the commercialized vitamin E-containing acetabular cup is normally less than 100 kGy.

More recently, some studies showed that the mechanical strength and fatigue resistance of vitamin E-stabilised UHMWPE can be further improved by some additional treatments. The further improvement of the wear resistance of vitamin E-stabilised irradiated UHMWPE was achieved by spatially manipulating the vitamin E concentration throughout the implant and limiting cross-linking to the surface [54–56]. Furthermore, the vitamin E-stabilised UHMWPE showed improved tensile and impact toughness and a good wear and oxidation resistance by high temperature melting after the introduction of vitamin E into UHMWPE [53].

3.3.2. Effect of Vitamin E on Tribological and Mechanical Properties of UHMWPE

Compared with conventional UHWMPE, vitamin E-stabilised UHMWPE has the ability to provide an improved tribological property [28]. Wear test results indicated that the wear rates of vitamin E-stabilised UHMWPE in clean serum and in serum with third-body particles had a 4-fold to 10-fold decrease from that of conventional UHMWPE [25]. The wear behaviour between vitamin E-stabilised UHMWPE and conventional UHMWPE using a knee joint simulator was compared, and the results indicated that vitamin E-stabilised UHMWPE had a consistently lower wear volume than that of the non-treated material [57]. It was believed vitamin E addition could prevent delamination caused by oxidation and fatigue, and therefore, significantly influence the life-span of the UHMWPE component.

The mechanical properties of vitamin E-stabilised UHMWPE are affected by vitamin E concentrations. The mechanical properties of vitamin E-stabilised UHMWPE are insensitive to the introduction of vitamin E up to the concentration of 0.4 wt% [51,52]. When vitamin E concentration is in a range of 0.01–0.05 wt%, vitamin E has a negligible effect on the mechanical properties of unsterilized and gamma sterilized UHMWPE [51]. Furthermore, for vitamin E concentration of 0 wt%, 0.1 wt%, 0.2 wt% and 0.4 wt%, the elastic modulus, elongation, tensile, and impact strength have no significant difference for both virgin and gamma sterilized UHMWPE [58]. However, for the 0.8 wt% vitamin E stabilised material, the elastic modulus and tensile strength were 20% lower than that of the virgin and gamma sterilized UHMWPE, while the elongation and impact strength remained at the baseline values [58]. To sum up, the previous studies [51,52,58] showed that vitamin E alone with up to 0.4 wt% concentration have a marginal effect on the mechanical properties of UHMWPE. However, some studies [44,57,59] suggested that vitamin E-stabilised UHMWPE has a significantly higher tensile strength, yield strength, elongation at break and fatigue resistance, especially at concentrations exceeding 0.1 wt% [59]. More detailed information on the mechanical properties of vitamin E-stabilised UHMWPE is shown in Table 2. Further studies on an optimal dose or an optimal dose range and better understanding of the mechanisms of the property improvement are needed.

3.3.3. Limitations and Challenges

Although vitamin E is an effective free radical scavenger, it impacts the crosslinking efficiency. Crosslinking radiation dose is an essential factor to the wear resistance. The introduction of vitamin E-stabilised material without any other changes in radiation exposure could result in a lower crosslinking density and has a potential trade off in wear resistance [40]. Therefore, it is a challenging work for orthopaedic implant researchers and manufacturers to optimise and determine an appropriate vitamin E concentration and radiation dose to achieve a good wear resistance.

3.4. *Fillers (CNFs, CNTs, Graphene and Hard Particles)*

A wide range of fillers are used to enhance the mechanical and lubrication properties of UHMWPE. Table 3 gives an overview of the commonly used filler materials and their effects on the mechanical properties and the wear performance. Details of the filler materials and their existing studies will be presented in the following sub-sections.

Table 2. The mechanical properties of vitamin E-stabilised UHMWPEs.

Resin	VE Concentration (wt%)	Tensile Strength Increase (%)	Strain Break Increase (%)	Yield Strength Increase (%)	Impact Strength Increase (%)	Elastic Modulus Increase (%)	Elongation Increase (%)	Fatigue Strength Increase (%)	Ref
UHMWPE GUR 1020 and 1050	0.01–0.05	No significant change	No significant change	-	-	-	-	-	[51]
UHMWPE GUR 1020	0.1–0.4	No significant change	-	-	No significant change	No significant change	No significant change	-	[58]
UHMWPE GUR 1020	0.8	−20	-	-	No significant change	−20	No significant change	-	[58]
UHMWPE GUR 1050	-	-	-	-	-	-	-	58	[60]
UHMWPE GUR 1050	0.1–0.3	51	-	19	-	-	−19	-	[52]
UHMWPE GUR 1050	-	36	-	4	-	-	32	35	[25]
UHMWPE	0.1–0.3	-	-	-	-	-	-	10–20	[59]

3.4.1. Carbon Nanofibres (CNFs)

Owing to the intrinsic properties and cytocompatibility of carbon nanofibres (CNF), incorporation of CNF to improve the mechanical and wear properties of UHMWPE receives increasing attention [61]. Studies on the mechanical properties of CNF reinforced polyethylene revealed that the material's yield stress, tensile modulus and hardness increased by the addition of a certain amount of CNFs. For example, it was reported that incorporation of 5 wt% CNFs resulted in an approximately 25% increase in the tensile modulus, while increasing the CNF content to 10 wt% led to a slightly additional benefit [62]. Similar trends were also observed for the hardness of CNF reinforced UHMWPE material [62]. The addition of the carbon nano-fibres results in an approximately linear increase in hardness. In addition, presence of the nano-fibre in the 10 wt% samples leads to an additional benefit in terms of a further increase in the hardness. The mechanical properties of CNF reinforced polyethylene are sensitive to the CNF concentration. In Sui's work [26], 0.5 wt%, 1.0 wt%, and 3.0 wt% CNF were added into UHMWPE. The tensile strength and tensile modulus of the UHMWPEs increased with the addition of CNFs, while a decrease appeared at 3 wt% content.

The presence of dispersed carbon nanofibres also can lead to a significantly improvement of wear performance of UHMWPE material. With the addition of 0.5–3 wt% CNFs, the wear rate of UHMWPEs decreased 56%–58% [26,60,63]. However, it was found that nanofiber agglomeration offset this enhancement. In Xu's study, the addition of 0.5 wt% CNF resulted in an improved wear resistance, while a higher level CNF addition (3 wt%) resulted in a higher wear rate [64]. A similar trend was found in Galetz's study [62], in which with 5 wt% CNF, the nanocomposite showed a highly improved wear rate compared to the pure UHMWPE. Compared with 5 wt% CNF-reinforced UHMWPE, the wear rate increased at the presence of a larger CNF concentration (10 wt%).

Table 3. Popularly used fillers and their effects on the mechanical properties and/or wear performance of UHMWPE.

Fillers	Percentage of Inclusion	Improved Properties	Reduction in Wear Rate
Carbon nanofibers (CNF) [26,58,59,62,65]	0.5%–5%	Tensile strength	56%–58%
Carbon nanotube (CNTs) [66,69]	0.1%–5%	Tensile strength Young's modulus Toughness	26%–86%
Graphene [70,75]	0.1%–1.0%	Lubrication, tensile strength Yield strength Reducing friction coefficient	2.5–4.5 times (depending on load)
Hard particles [76,79]	10%–20%	Bearing loading capacity	36%–60%

3.4.2. Carbon Nanotube (CNTs)

The introduction of carbon nanotubes (CNT) is a recent development that will likely make a new generation of advanced implant material possible. CNTs possess high strength, a high aspect ratio, and excellent thermal and electrical conductivity [80]. They can enhance the mechanical performance and wear resistance of UHMWPE, promote bone growth and act as an antioxidant.

Experimental investigations on the mechanical properties of CNT reinforced polyethylene indicated that the tensile strength, Young's modulus and toughness of the material increased by the addition of a certain amount of CNTs [72,81]. In previous studies [81–84], with the addition of 0.1–5 wt% CNTs, the tensile strength and Young's modulus of reinforced UHMWPE increased 8%–38% and 5%–100%, respectively. An improvement in the UHMWPE toughness was also reported with the addiction of 1 wt% CNTs [68].

In addition to the improved mechanical properties, CNT-reinforced UHMWPE has gained intensive interest because of their good wear resistance. By the incorporation of 0.1–4 wt% CNTs, the wear rate of UHMWPEs decreased 26%–86% [69,80,81,83,85]. However, the wear performance of CNT-reinforced UHMWPE is sensitive to CNT concentrations and types. The friction coefficient of pure UHMWPE is approximately 0.05, while the friction coefficient increased slowly as CNTs were added from 0.1 to 0.5 wt% [80]. The weight loss of CNT-reinforced UHMWP was found to decrease with the increasing amount of CNT addition [80]. Furthermore, in Liu's study [86], 0.45 wt% multi-walled carbon nanotubes (MWCNTs), 0.1 wt% granular form carbon nanotubes and 0.5 wt% single-walled carbon nanotubes were blended into UHMWPE. The wear rates of these three types of CNT reinforced UHMWPEs were slightly higher than that of pure UHMWPE material.

Martinez-Morales *et al.* [87] reported that MWCNTs improve the oxidative resistance of the composite (MWCNT/UHMWPE) and some *in vitro* studies showed that bone cells could grow along CNTs [88]. It is an exciting prospect for acetabular sockets as the structured surface of CNTs composite could potentially encourage new bone growth on the outer surface of the prosthesis so that the bond between natural bone and prosthesis will be strengthened. Last but not least, CNTs can improve the antioxidant ability in the polymer matrix because of their radical accepting capacity [73].

3.4.3. Graphene

The outstanding mechanical and physical properties of graphene make this material a candidate to be used in UHMWPE composite as a component of AJs. This two-dimensional (2D) material has excellent mechanical properties, such as the Young modulus (0.5–1 Tpa), tensile strength (~130 Gpa) and high thermal and electrical conductivity (~4.84 $\times$ 10^3 W/mk and 7600 S/m, respectively) [70,74]. These outstanding properties made this material suitable for different applications, including computing, nano and micro-electronical systems, and biomedical fields.

Graphene is produced using different methods. The graphene nanoplatelet (GNP) is obtained by the electrostatic spraying technique. The graphene oxide (GO) is obtained by treating the graphite with a strong oxidazer [71], resulting in a compound of carbon, oxygen and hydrogen. The reduced oxide graphene (RGO) is a result from chemical exfoliation of graphene [75]. The properties of graphene are affected by the employed method of acquisition.

Graphene has been added as a filling material in UHMWPE composite to improve the wear resistance of the UHMWPE component used for artificial implants. Previous studies have shown an improvement in the wear resistance of this component with an increase in the graphene content [72]. Lahiri *et al.* [77] evaluated the evolution of the wear resistance at the nano-scale by scratching the UHMWPE-GNP composite at different GNP concentrations (0.1, 0.5, and 1.0 wt%) and using three different loads (100, 200 and 300 μN). The authors noticed that the wear resistance of the material improve by 4.5 times with a 1 wt% GNP, compared with the pure UHMWPE. The coefficient of friction (COF) did not report a significant increase for a higher GNP content, which may be associated to the released graphene on the scratch, lubricating the path and reducing the COF.

3.4.4. Hard Particles

In addition to CNFs and CNTs, particle reinforcements using other hard particles were reported to be a very promising method for an improvement in the wear performance of UHMWPE [78,79,89–91]. By means of incorporating Ti particles [89], Zirconium particles [78], Pt-Zr quasicrystal [79], natural coral particles [90], and quartz [91] into UHMWPE, the wear rates of the particle-reinforced UHMWPEs reduced from 36% to 60%. The mechanism of this improvement is that these hard particles can effectively take bearing load to protect the polymer matrix and thus improve the wear resistance of UHMWPE. However, the worn off hard particles might also act as a third body in the tribological system to accelerate the wear rate [92].

3.4.5. Limitations and Challenges

The effective use of carbon nanofillers in UHMWPE depends upon its homogeneous dispersion throughout the polymer matrix. Due to the complexity of UHMWPE material, the incorporation of CNF or CNT into UHMWPE is greatly complicated by the extremely high viscosity of the polymer matrix. The agglomerations of carbon nanofillers with a size of several tens of microns could act as localized stress concentration points [62]. These weak points could be removed from the bulk material in a wear process, resulting in wear particles. The wear debris could act as a third body between the interfaces and accelerate the material degradation process.

The mechanical properties of hard particles, such as hardness, size and concentration, are essential factors to the wear performance of particle-reinforced UHMWPE. Further study is necessary to determine an optimal particle type, particle size and concentration to minimise or eliminate the side effects of the incorporated particles on the wear performance of UHMWPE material. It is also important to study the interaction of wear debris generated from the particle reinforced UHMWPE with surrounding tissues to evaluate and eliminate possible inflammatory reactions, which may cause osteolysis.

In summary, the key challenges of carbon nanofiller reinforced UHMNWPE for tribological application is the determination of type of carbon nanofiller, the optimisation of carbon nanofiller concentration, and the improvement of blending techniques. Furthermore, knowledge in the public domain about carbon nanofillers reinforced UHMWPE for orthopaedic prostheses remains in its infancy. Additional research is needed to comprehensively study the tribological properties and to warrant its biocompatibility.

3.5. Surface Modification of UHMWPE

When used to replace damaged or diseased joints, UHMWPE made joints interact with many different components such as hard bone, soft tissue, blood, and synovial fluid. Similar to other biomaterials, whose functionality is significantly influenced by the surface characteristics, the surface structures and textures of UHMWPE determine its biocompatibility, physicochemical surface properties (e.g., wettability, surface energy), mechanical properties (e.g., toughness and load-carrying capacity) and tribological performance. A large amount of studies were carried out on surface modification of UHMWPE to improve the function and lifetime of the artificial joints. This section reviews commonly used surface modification methods, in particular, surface texturing techniques for an improvement in the mechanical properties and wear resistance of UHMWPE materials.

3.5.1. Surface Coating

Coating techniques involve depositing a thin film on the base material or promoting the cross-linking reaction by changing the properties of the outmost surface which is reviewed in Section 3.2. They are used popularly to enhance the performance of biomaterials. A range of properties can be improved including osteoinductive properties [93], hydrophilicity [94], mechanical and biological properties [95,96], surface properties (e.g., surface free energy) [97], and biotribological properties [94]. Surface coating by adding materials on the top surface will be briefly reviewed in the next paragraph.

Commonly applied coating techniques include plasma-surface modification using plasma spraying, implantation or deposition techniques [98] to enhance the surface properties such as wear resistance, and hardness. In the process, often a thin layer of a hard metallic material is added on the bulk material to improve its properties. More recently, nylon coating [95] on UHMWPE was reported to achieve lower cytotoxicity, less wear debris-induced osteolysis, and superior mechanical properties compared to neat UHMWPE. Zirconium carbon nitrides (ZrC_xN_{1-x}) coatings with embedded silver nanoparticles [94] were used as a surface modification option for multifunctional purposes including antimicrobial effect, corrosion resistance and maintaining or even improving their mechanical and tribological performance. Hydrogenated diamond like carbon (DLCH) coating has been also tested on UHMWPE with a good performance on the wear resistance, hardness and biocompatibility [99]. Two groups of samples were tested at a sliding velocity of 0.05 m/s and a load of 5.23 N. The study reported a wear rate reduction of approximately 30% when the samples were surface coated with a thin layer (<20 nm) of DLCH. Nowadays, with the advancement of nanotechnology, nano-materials such as nano carbon tubes [96], have been used as a coating material to further improve the performance of the coating techniques.

3.5.2. Ion Beam Surface Modification

Ion implantation is used to generate UHMWPE surface layers that are integrated with the substrate and have a specified structure or composition. The ion-beam texturing process can be performed to produce either a square or a round pattern using an electron formed screen mesh mask to the surface. Compared with other coating methods, the advantages of ion beam surface modification are that this technique avoids the risk of delamination and that the process is extremely controllable and reproducible [100]. Improvement of the tribological behaviour, as well as the surface mechanical properties of polymeric materials, was reported in [101] and surface wear reduction was found on

UHMWPE by using nitrogen ion implantation [102,103]. It is believed that this surface modification method has hardening and stiffening effects. It can also increase the wettability of the surface. Dong *et al.* reported a wear rate reduction from 1×10^{-8} $mm^3N^{-1}m^{-1}$ for the untreated samples to 0.1 $\times 10^{-8} mm^3N^{-1}m^{-1}$ for the PI^3 treated ones [12]. Traditional ion implantation technique involves line-of-sight processing and thus is not suitable for the treatment of bone implants with a complex shape and curved surface [104]. To overcome the limitation, plasma immersion ion implantation is used for improvement of the surface mechanical properties, biocompatibility, bioactivity, antibacterial activity and the wear performance of biomaterials. Nitrogen, oxygen and hydrogen implanted UHMWPE showed better wear resistance than those untreated [105,106].

3.5.3. Photolithography and Nanoimprint Lithography

Photolithgraphy makes surface patterns with a mask. The process involves coating, etching and transferring a photo-resist layer to the surface of a bulk material whose surface properties need to be improved [103]. In the study conducted in [107], photolithograpy was used to generate dimples with a diameter of 50 μm, a depth of 15 μm and the area density in a range of 5% to 40% on both stainless steel and UHMWPE surfaces. The study found that having dimples on the UHMWPE surface was more effective for friction reduction than on a stainless steel surface. At high load (700 N), only the patterned UHMWPE surface could effectively reduce friction.

With advances in 3D and nano technology, nanoimprint lithograph (NIL) has become a versatile technique to produce nano-patterns on the surface of UHMWPE. The technique has many attractive features including nano-scale resolution, high throughput, high controllability of 3D nanostructures, low cost, and ability of generating patterns over a large area [100]. NIL can generate features with a few nano-metre resolution on either flat or nonflat surfaces. It is used to improve the tribological properties of the surface of UHMWPE. In the study conducted in [108], the patterned UHMWPE surface had a reduction in the friction coefficient between 8% and 35% in comparison to the non-patterned surface when tested under a dry sliding condition and with normal load in a range of 60 to 200 mN.

3.5.4. Laser Surface Texturing

Texturing or surface engineering refers to the physical modification of the surface morphology, and is now accepted as an effective and feasible technique for friction reduction and improvement of wear resistance. Due to the interaction of two moving surfaces occurring at various scales from macro to nano scale [109], research on surface texturing at the micro- and nano-metre scale has recently been carried out with the majority of existing studies being in the micron scale.

Commonly used surface texturing techniques for UHMWPE are electrical discharge etching [110] and laser surface engineering [111]. Laser micro-machining is now widely used to modify the 3D surface textures of UHMWPE at both micro- and submicro-metre scales. This technique becomes a good alternative to a standard photolithographic process because it is rapid and clean [111]. Using experimental and/or simulation methods existing studies investigated the effects of dimple size, depth and area density on the friction and wear properties of UHMWPE and its counterpart.

One common outcome is that the micro-texture increases the loading-carrying capacity and the thickness of the joint lubricant film, resulting in a reduction in friction and wear [110,112–115]. Load effects were studied in [107,113] when dimple diameter was around 40–50 μm. It was found that under a light load (17–100 N), surface textures on either UHMWPE or its steel counterpart reduced friction. At high load (e.g., 700 N in [107]), it was reported that only the textures on the UHMWPE surface could effectively reduce friction. Wang *et al.* [115] investigated the area ratio effects and found that an area ratio in a range of 20%–40% was ideal to maximise the total hydrodynamic pressure between the two moving surfaces. They also found that when dimples were fabricated on the surface of steel and at a light load, the area density of 5%–15% was effective for friction reduction. If the dimples were textured on the UHMWPE surface, the area density of 30% improved both friction and wear resistance under a water lubricated condition and at a relative high load. It is believed that by applying surface textures on either UHMWPE or steel surface, abrasive wear is reduced by improving the lubrication condition through storing wear particles and lubricant in dimples [110] and by reduction of frictional heating of the sliding surfaces [114].

3.5.5. Limitations and Challenges

Possibly due to the capability of current laser technology, only micro-sized dimples were fabricated on UHMWPE surfaces and their effects were investigated in existing studies. Other patterns, such as square, ellipse, need to be fabricated and examined to find an optimal surface texture for friction and wear reduction. Furthermore, nano-scale surface textures were reported to provide low friction and wear properties [108] and need to be studied further.

Modifying surface texture to improve the wear performance and life span of the material is a challenging and debatable topic. Although some encouraging outcomes have been achieved in the past as reviewed above, its long term effect needs further investigations. For example, Kelly *et al.* [116] evaluated the wear damage of the tibiofemoral articular surface and the mobile-bearing surface (refer to Figure 2A–D of [116]). It was reported that the UHMWPE components became degraded in a relative short time (three years) and the most predominant damages modes were scratching, burnishing and pitting. The scratches, in particular, were believed to be made by a third body wear caused by particles from surgery or particles of the bone cement. Due to the presence of wear debris generated in the wear process and other particles, the wear process and mechanism are often evolved with the surface texture. The initially designed surface morphologies can be modified in the wear process. The long term, practical effect of the surface modification approaches requires further studies.

4. Conclusions

The ability to replace diseased or traumatised natural joints with artificial prostheses in order to alleviate pain and disability has been one of the major successes in engineering and medicine over the last 50 years. Although UHMWPE has been used as the most popular artificial replacement material in clinical applications for five decades, continuous research has been conducted to improve the performance of the material so its service life can be further extended. This paper has reviewed reported methods for improving the mechanical properties through creating a cross-linked structure, using an irradiation process with and without vitamin E, and by adding micron or nano-particles. Surface engineering techniques for the improvement of the lubrication conditions are also reviewed. Existing studies have demonstrated that cross-linked UHMWPE by gamma radiation is an effective way to significantly improve the wear resistance of the material from a wear rate of about 20 mm^3 per million cycles of untreated material to close to a zero wear rate for gamma radiated to 280 kGy. Including fillers can also improve the mechanical properties and wear resistance. However, the reported improvement varies from 25% to 86% depending on different filler materials, their concentrations and/or directions. Similarly, according to the reported results, surface modifications can improve the wear resistance from 30% for the surface coating of diamond like carbon to 90% using ion beam surface modification method. As wear is a material removal process, the long term effect of the surface texture modification methods need to be further studied. With advances in modern technology, it is anticipated that multi-approaches will be developed and used to improve the chain structure, to further strengthen the mechanical properties and to improve the lubrication condition simultaneously in future studies.

Author Contributions: Juan C. Baena drafted the mechanical properties and manufacturing process of UHMWPE section. He also collected and presented the materials in Sections 3.1 3.2 of the paper. Jingping Wu drafted Section 3.3 and 3.4 while Zhongxiao Peng was responsible for Section 3.5 and the overall flow and quality of this paper. The three authors worked together to draft the other parts of the paper including Abstract, Introduction and Conclusions.

Conflicts of Interest: The authors declare no conflict of interest.

References

1. Kanaga Karuppiah, K.S.; Bruck, A.L.; Sundararajan, S.; Wang, J.; Lin, Z.; Xu, Z.-H.; Li, X. Friction and wear behavior of ultra-high molecular weight polyethylene as a function of polymer crystallinity. *Acta Biomater.* **2008**, *4*, 1401–1410. [CrossRef] [PubMed]
2. Garellick, G.; Maichou, H.; Herberts, P. Specific or general health outcome measures in the evaluation of total hip replacement. *J. Bone Joint Surg.* **1998**, *80*, 600–606. [CrossRef]
3. Wayne, G. *A National Public Health Agenda for Osteoarthritis*; Centers for Disease Control and Prevention: DeKalb County, GA, USA, 2010.
4. Delport, H.P.; Banks, S.A.; De Schepper, J.; Bellemans, J. A kinematic comparison of fixed- and mobile-bearing knee replacements. *J. Bone Joint Surg. Br.* **2006**, *88*, 1016–1021. [CrossRef] [PubMed]
5. Williams, J.A. Wear and wear particles—Some fundamentals. *Tribol. Int.* **2005**, *38*, 863–870. [CrossRef]
6. Jin, Z.; Fisher, J. 2—Tribology in joint replacement. In *Joint Replacement Technology*; Revell, P.A., Ed.; Woodhead Publishing: Sawston, UK, 2008; pp. 31–55.
7. Bono, J.V.; Scott, R.D. *Revision Total Knee Arthroplasty*; Springer: Berlin/Heidelberg, Germany, 2005; p. 292.
8. Bhushan, B. Adhesion and stiction: Mechanisms, measurement techniques, and methods for reduction. *J. Vac. Sci. Technol. B* **2003**, *21*. [CrossRef]
9. Jacobs, J.J.; Skipor, A.K.; Patterson, L.M.; Hallab, N.J.; Paprosky, W.G.; Black, J.; Galante, J.O. Metal release in patients who have had a primary total hip arthroplasty. A prospective, controlled, longitudinal study. *J. Bone Joint Surg. Am.* **1998**, *80*, 1447–1458. [PubMed]
10. Shi, W.; Li, X.Y.; Dong, H. Preliminary investigation into the load bearing capacity of ion beam surface modified UHMWPE. *J. Mater. Sci.* **2004**, *39*, 3183–3186. [CrossRef]
11. Pearle, A.D.; Warren, R.F.; Rodeo, S.A. Basic science of articular cartilage and osteoarthritis. *Clin. Sports Med.* **2005**, *24*, 1–12. [CrossRef] [PubMed]
12. Dong, H.; Shi, W.; Bell, T. Potential of improving tribological performance of UHMWPE by engineering the Ti6Al4V counterfaces. *Wear* **1999**, *225–229*, 146–153. [CrossRef]
13. Deng, Y.; Xiong, D.; Wang, K. Biotribological properties of UHMWPE grafted with AA under lubrication as artificial joint. *J. Mater. Sci. Mater. Med.* **2013**, *24*, 2085–2091. [CrossRef] [PubMed]
14. Liao, Y.; Hoffman, E.; Wimmer, M.; Fischer, A.; Jacobs, J.; Marks, L. CoCrMo metal-on-metal hip replacements. *Phys. Chem. Chem. Phys.* **2013**, *15*. [CrossRef] [PubMed]
15. Zhang, H.-X.; Shin, Y.-J.; Lee, D.-H.; Yoon, K.-B. Preparation of ultra high molecular weight polyethylene with $MgCl_2/TiCl_4$ catalyst: Effect of internal and external donor on molecular weight and molecular weight distribution. *Polymer Bull.* **2011**, *66*, 627–635. [CrossRef]
16. Kurtz, S.M. Chapter 1—A primer on uhmwpe. In *UHMWPE Biomaterials Handbook*, 2nd ed.; Kurtz, S.M., Ed.; Academic Press: Boston, MA, USA, 2009; pp. 1–6.
17. Edidin, A.A.; Kurtz, S.M. Influence of mechanical behavior on the wear of 4 clinically relevant polymeric biomaterials in a HIP simulator. *J. Arthroplast.* **2000**, *15*, 321–331. [CrossRef]
18. Brach del Prever, E.M.; Bistolfi, A.; Bracco, P.; Costa, L. UHMWPE for arthroplasty: Past or future? *J. Orthop. Traumatol.* **2009**, *10*, 1–8. [CrossRef] [PubMed]
19. Kurtz, S.M. Chapter 2—From ethylene gas to uhmwpe component: The process of producing orthopedic implants. In *The UHMWPE Handbook*; Kurtz, S.M., Ed.; Academic Press: San Diego, CA, USA, 2004; pp. 13–36.
20. Lim, K.L.K.; Ishak, Z.A.M.; Ishiaku, U.S.; Fuad, A.M.Y.; Yusof, A.H.; Czigany, T.; Pukanszky, B.; Ogunniyi, D.S. High-density polyethylene/ultrahigh-molecular-weight polyethylene blend. I. The processing, thermal, and mechanical properties. *J. Appl. Polymer Sci.* **2005**, *97*, 413–425. [CrossRef]
21. Kurtz, S.M. Chapter 2—From ethylene gas to UHMWPE component: The process of producing orthopedic implants. In *UHMWPE Biomaterials Handbook*, 2nd ed.; Kurtz, S.M., Ed.; Academic Press: Boston, MA, USA, 2009; pp. 7–19.
22. Ayache, J.; Beaunier, L.; Boumendil, J.; Ehret, G.; Laub, D. *Introduction to Materials*; Springer: Berlin/Heidelberg, Germany, 2010; pp. 3–31.
23. McKeen, L.W. 3—Plastics used in medical devices. In *Handbook of Polymer Applications in Medicine and Medical Devices*; Ebnesajjad, K.M., Ed.; William Andrew Publishing: Oxford, UK, 2014; pp. 21–53.
24. Wang, S.; Ge, S. The mechanical property and tribological behavior of UHMWPE: Effect of molding pressure. *Wear* **2007**, *263*, 949–956. [CrossRef]

25. Oral, E.; Christensen, S.D.; Malhi, A.S.; Wannomae, K.K.; Muratoglu, O.K. Wear resistance and mechanical properties of highly cross-linked, ultrahigh-molecular weight polyethylene doped with vitamin E. *J. Arthroplasty* **2006**, *21*, 580–591. [CrossRef] [PubMed]
26. Sui, G.; Zhong, W.H.; Ren, X.; Wang, X.Q.; Yang, X.P. Structure, mechanical properties and friction behavior of UHMWPE/HDPE/carbon nanofibers. *Mater. Chem. Phys.* **2009**, *115*, 404–412. [CrossRef]
27. Medel, F.J.; Puertolas, J.A. Wear resistance of highly cross-linked and remelted polyethylenes after ion implantation and accelerated ageing. *Proc. Inst. Mech. Eng. Part H J. Eng. Med.* **2008**, *222*, 877–885. [CrossRef]
28. Oral, E.; Godleski Beckos, C.A.; Lozynsky, A.J.; Malhi, A.S.; Muratoglu, O.K. Improved resistance to wear and fatigue fracture in high pressure crystallized vitamin E-containing ultra-high molecular weight polyethylene. *Biomaterials* **2009**, *30*, 1870–1880. [CrossRef] [PubMed]
29. Jahan, M.S. Chapter 29—Esr insights into macroradicals in uhmwpe. In *Uhmwpe Biomaterials Handbook*, 2nd ed.; Kurtz, S.M., Ed.; Academic Press: Boston, MA, USA, 2009; pp. 433–450.
30. Green, T.R.; Fisher, J.; Stone, M.H.; Wroblewski, B.M.; Ingham, E. Polyethylene particles of a 'critical size' are necessary for the induction of cytokines by macrophages in vitro. *Biomaterials* **1998**, *19*, 2297–2302. [CrossRef] [PubMed]
31. Muratoglu, O.K. Chapter 13—Highly crosslinked and melted UHMWPE. In *UHMWPE Biomaterials Handbook*, 2nd ed.; Kurtz, S.M., Ed.; Academic Press: Boston, MA, USA, 2009; pp. 197–204.
32. McKellop, H.; Shen, F.W.; Lu, B.; Campbell, P.; Salovey, R. Development of an extremely wear-resistant ultra high molecular weight polyethylene for total hip replacements. *J. Orthop. Res.* **1999**, *17*, 157–167. [CrossRef] [PubMed]
33. Furmanski, J.; Anderson, M.; Bal, S.; Greenwald, A.S.; Halley, D.; Penenberg, B.; Ries, M.; Pruitt, L. Clinical fracture of cross-linked UHMWPE acetabular liners. *Biomaterials* **2009**, *30*, 5572–5582. [CrossRef] [PubMed]
34. Costa, L.; Luda, M.P.; Trossarelli, L. Ultra-high molecular weight polyethylene: I. Mechano-oxidative degradation. *Polymer Degrad. Stab.* **1997**, *55*, 329–338. [CrossRef]
35. Costa, L.; Luda, M.P.; Trossarelli, L.; Brach del Prever, E.M.; Crova, M.; Gallinaro, P. *In vivo* uhmwpe biodegradation of retrieved prosthesis. *Biomaterials* **1998**, *19*, 1371–1385. [CrossRef] [PubMed]
36. Premnath, V.; Harris, W.H.; Jasty, M.; Merrill, E.W. Gamma sterilization of UHMWPE articular implants: An analysis of the oxidation problem. *Biomaterials* **1996**, *17*, 1741–1753. [CrossRef] [PubMed]
37. Stea, S.; Antonietti, B.; Baruffaldi, F.; Visentin, M.; Bordini, B.; Sudanese, A.; Toni, A. Behavior of hylamer polyethylene in hip arthroplasty: Comparison of two gamma sterilization techniques. *Int. Orthop.* **2006**, *30*, 35–38. [CrossRef] [PubMed]
38. Oral, E.; Malhi, A.S.; Muratoglu, O.K. Mechanisms of decrease in fatigue crack propagation resistance in irradiated and melted UHMWPE. *Biomaterials* **2006**, *27*, 917–925. [CrossRef] [PubMed]
39. Bracco, P.; Oral, E. Vitamin E-stabilized UHMWPE for total joint implants: A review. *Clin. Orthop. Relat. Res.* **2011**, *469*, 2286–2293. [CrossRef] [PubMed]
40. Kurtz, S.M. Chapter 16 vitamin-E-blended UHMWPE biomaterials. In *UHMWPE Biomaterials Handbook*, 2nd ed.; Academic Press: Boston, MA, USA, 2009; pp. 1–6.
41. Bracco, P.; Brunella, V.; Zanetti, M.; Luda, M.P.; Costa, L. Stabilisation of ultra-high molecular weight polyethylene with vitamin E. *Polymer Degrad. Stab.* **2007**, *92*, 2155–2162. [CrossRef]
42. Shibata, N.; Tomita, N.; Onmori, N.; Kato, K.; Ikeuchi, K. Defect initiation at subsurface grain boundary as a precursor of delamination in ultrahigh molecular weight polyethylene. *J. Biomed. Mater. Res. Part A* **2003**, *67A*, 276–284. [CrossRef]
43. Kiyose, C.; Ueda, T. Vitamin E distribution and metabolism of tocopherols and tocotrienols *in vivo*. *J. Clin. Biochem. Nutr.* **2004**, *35*, 47–52. [CrossRef]
44. Oral, E.; Wannomae, K.K.; Hawkins, N.; Harris, W.H.; Muratoglu, O.K. A-tocopherol-doped irradiated UHMWPE for high fatigue resistance and low wear. *Biomaterials* **2004**, *25*, 5515–5522. [CrossRef] [PubMed]
45. Oral, E.; Wannomae, K.K.; Rowell, S.L.; Muratoglu, O.K. Diffusion of vitamin E in ultra-high molecular weight polyethylene. *Biomaterials* **2007**, *28*, 5225–5237. [CrossRef] [PubMed]
46. Oral, E.; Muratoglu, O. Vitamin E diffused, highly crosslinked UHMWPE: A review. *Int. Orthop. (SICOT)* **2011**, *35*, 215–223. [CrossRef]
47. Oral, E.; Rowell, S.L.; Muratoglu, O.K. The effect of α-tocopherol on the oxidation and free radical decay in irradiated uhmwpe. *Biomaterials* **2006**, *27*, 5580–5587. [CrossRef] [PubMed]

48. Wannomae, K.K.; Christensen, S.D.; Micheli, B.R.; Rowell, S.L.; Schroeder, D.W.; Muratoglu, O.K. Delamination and adhesive wear behavior of α-tocopherol-stabilized irradiated ultrahigh-molecular-weight polyethylene. *J. Arthroplasty* **2010**, *25*, 635–643. [CrossRef] [PubMed]
49. Haider, H.; Weisenburger, J.N.; Kurtz, S.M.; Rimnac, C.M.; Freedman, J.; Schroeder, D.W.; Garvin, K.L. Does vitamin E-stabilized ultrahigh-molecular-weight polyethylene address concerns of cross-linked polyethylene in total knee arthroplasty? *J. Arthroplasty* **2012**, *27*, 461–469. [CrossRef] [PubMed]
50. Lerf, R.; Zurbrügg, D.; Delfosse, D. Use of vitamin e to protect cross-linked uhmwpe from oxidation. *Biomaterials* **2010**, *31*, 3643–3648. [CrossRef] [PubMed]
51. Kurtz, S.M.; Dumbleton, J.; Siskey, R.S.; Wang, A.; Manley, M. Trace concentrations of vitamin E protect radiation crosslinked UHMWPE from oxidative degradation. *J. Biomed. Mater. Res. Part A* **2009**, *90A*, 549–563. [CrossRef]
52. Oral, E.; Greenbaum, E.S.; Malhi, A.S.; Harris, W.H.; Muratoglu, O.K. Characterization of irradiated blends of alpha-tocopherol and UHMWPE. *Biomaterials* **2005**, *26*, 6657–6663. [CrossRef] [PubMed]
53. Fu, J.; Doshi, B.N.; Oral, E.; Muratoglu, O.K. High temperature melted, radiation cross-linked, vitamin E stabilized oxidation resistant UHMWPE with low wear and high impact strength. *Polymer* **2013**, *54*, 199–209. [CrossRef]
54. Oral, E.; Godleski Beckos, C.; Malhi, A.S.; Muratoglu, O.K. The effects of high dose irradiation on the cross-linking of vitamin E-blended ultrahigh molecular weight polyethylene. *Biomaterials* **2008**, *29*, 3557–3560. [CrossRef] [PubMed]
55. Muratoglu, O.K.; O'Connor, D.O.; Bragdon, C.R.; Delaney, J.; Jasty, M.; Harris, W.H.; Merrill, E.; Venugopalan, P. Gradient crosslinking of uhmwpe using irradiation in molten state for total joint arthroplasty. *Biomaterials* **2002**, *23*, 717–724. [CrossRef] [PubMed]
56. Oral, E.; Neils, A.; Muratoglu, O.K. High vitamin E content, impact resistant UHMWPE blend without loss of wear resistance. *J. Biomed. Mater. Res. Part B* **2015**, *103*, 790–797. [CrossRef]
57. Teramura, S.; Sakoda, H.; Terao, T.; Endo, M.M.; Fujiwara, K.; Tomita, N. Reduction of wear volume from ultrahigh molecular weight polyethylene knee components by the addition of vitamin E. *J. Orthop. Res.* **2008**, *26*, 460–464. [CrossRef] [PubMed]
58. Wolf, C.; Krivec, T.; Lederer, K.; Schneider, W. Examination of the suitability of alpha-tocopherol as a stabilizer for ultra-high molecular weight polyethylene used for articulating surfaces in joint endoprostheses. *J. Mater. Sci. Mater. Med.* **2002**, *13*, 185–189. [CrossRef] [PubMed]
59. Tomita, N.; Kitakura, T.; Onmori, N.; Ikada, Y.; Aoyama, E. Prevention of fatigue cracks in ultrahigh molecular weight polyethylene joint components by the addition of vitamin E. *J. Biomed. Mater. Res.* **1999**, *48*, 474–478. [CrossRef] [PubMed]
60. Wood, W.J.; M, R.G.; Zhong, W.H. Improved wear and mechanical properties of UHMWPE-carbon nanofiber composites through an optimized paraffin-assisted melt-mixing process. *Compos. Part B* **2011**, *42*, 584–891. [CrossRef]
61. Rachel, L.; Price, K.A.T.J.W. Improved osteoblast viability in the presence of smaller nanometre dimensioned carbon fibres. *Nanotechnology* **2004**, *15*, 892–900. [CrossRef]
62. Galetz, M.C.; Blaβ, T.; Ruckdäschel, H.; Sandler, J.K.W.; Altstädt, V.; Glatzel, U. Carbon nanofibre-reinforced ultrahigh molecular weight polyethylene for tribological applications. *J. Appl. Polymer Sci.* **2007**, *104*, 4173–4181. [CrossRef]
63. Wood, B.L.; Zhong, W. Influence of phase morphology on the sliding wear of polyethylene blends filled with carbon nanofibers. *Polymer Eng. Sci.* **2010**, *50*, 613–623. [CrossRef]
64. Xu, S.; Aydar, A.; Liu, T.; Weston, W.; Tangpong, X.W.; Akhatov, I.S.; Zhong, W.-H. Wear of carbon nanofiber reinforced HDPE nanocomposites under dry sliding condition. *J. Nanotechnol. Eng. Med.* **2012**, *3*, 041003. [CrossRef]
65. Price, R.L.; Haberstroh, K.M.; Webster, T.J. Improved osteoblast viability in the presence of smaller nanometre dimensioned carbon fibres. *Nanotechnology* **2004**, *15*, 892–900. [CrossRef]
66. Bakshi, S.R.; Tercero, J.E.; Agarwal, A. Synthesis and characterization of multiwalled carbon nanotube reinforced ultra high molecular weight polyethylene composite by electrostatic spraying technique. *Compos. Part A* **2007**, *38*, 2493–2499. [CrossRef]
67. Campo, N.; Visco, A.M. Incorporation of carbon nanotubes into ultra high molecular weight polyethylene by high energy ball milling. *Int. J. Polymer Anal. Charact.* **2010**, *15*, 438–449. [CrossRef]

68. Ruan, S.L.; Gao, P.; Yang, X.G.; Yu, T.X. Toughening high performance ultrahigh molecular weight polyethylene using multiwalled carbon nanotubes. *Polymer* **2003**, *44*, 5643–5654. [CrossRef]
69. Xue, Y.; Wu, W.; Jacobs, O.; Schadel, B. Tribological behaviour of UHMWPE/HDPE blends reinforced with multi-wall carbon nanotubes. *Polymer Test.* **2006**, *25*, 221–229. [CrossRef]
70. Kuilla, T.; Bhadra, S.; Yao, D.; Kim, N.H.; Bose, S.; Lee, J.H. Recent advances in graphene based polymer composites. *Prog. Polymer Sci.* **2010**, *35*, 1350–1375. [CrossRef]
71. Min, C.; Nie, P.; Song, H.-J.; Zhang, Z.; Zhao, K. Study of tribological properties of polyimide/graphene oxide nanocomposite films under seawater-lubricated condition. *Tribol. Int.* **2014**, *80*, 131–140. [CrossRef]
72. Puértolas, J.A.; Kurtz, S.M. Evaluation of carbon nanotubes and graphene as reinforcements for UHMWPE-based composites in arthroplastic applications: A review. *J. Mech. Behav. Biomed. Mater.* **2014**, *39*, 129–145. [CrossRef] [PubMed]
73. Shi, X.; Wang, J.; Jiang, B.; Yang, Y. Hindered phenol grafted carbon nanotubes for enhanced thermal oxidative stability of polyethylene. *Polymer* **2013**, *54*, 1167–1176. [CrossRef]
74. Soldano, C.; Mahmood, A.; Dujardin, E. Production, properties and potential of graphene. *Carbon* **2010**, *48*, 2127–2150. [CrossRef]
75. Stankovich, S.; Dikin, D.A.; Piner, R.D.; Kohlhaas, K.A.; Kleinhammes, A.; Jia, Y.; Wu, Y.; Nguyen, S.T.; Ruoff, R.S. Synthesis of graphene-based nanosheets via chemical reduction of exfoliated graphite oxide. *Carbon* **2007**, *45*, 1558–1565. [CrossRef]
76. Ge, S.; Zhang, D.; Wang, Q. Biotribological behaviour of ultra-high molecular weight polyethylene composites containing Ti in a hip joint simulator. *Proc. Inst. Mech. Eng. Part J.* **2007**, *221*, 307–313.
77. Lahiri, D.; Hec, F.; Thiesse, M.; Durygin, A.; Zhang, C.; Agarwal, A. Nanotribological behavior of graphene nanoplatelet reinforced ultra high molecular weight polyethylene composites. *Tribol. Int.* **2014**, *70*, 165–169. [CrossRef]
78. Plumlee, K.; Schwartz, C.J. Improved wear resistance of orthopaedic UHMWPE by reinforcement with zirconium particles. *Wear* **2009**, *267*, 710–717. [CrossRef]
79. Schwartz, C.J.; Bahadur, S.; Mallapragada, S.K. Effect of crosslinking and Pt-Zr quasicrystal fillers on the mechanical properties and wear resistance of uhmwpe for use in artificial joints. *Wear* **2007**, *263*, 1072–1080. [CrossRef]
80. Zoo, Y.-S.; An, J.-W.; Lim, D.-P.; Lim, D.-S. Effect of carbon nanotube addition on tribological behavior of UHMWPE. *Tribol. Lett.* **2004**, *16*, 305–309. [CrossRef]
81. Kanagaraj, S.; Mathew, M.T.; Fonseca, A.; Oliveira, M.S.A.; Simões, J.A.O.; Rocha, L.A. Tribological characterisation of carbon nanotubes/ultrahigh molecular weight polyethylene composites: The effect of sliding distance. *Int. J. Surf. Sci. Eng.* **2010**, *4*, 305–321. [CrossRef]
82. Meschi Amoli, B.; Ahmad Ramazani, S.A.; Izadi, H. Preparation of ultrahigh-molecular-weight polyethylene/carbon nanotube nanocomposites with a ziegler-Natta catalytic system and investigation of their thermal and mechanical properties. *J. Appl. Polym. Sci.* **2012**, *125*, E453–E461.
83. Maksimkin, A.V.; Kaloshkin, S.D.; Kaloshkina, M.S.; Gorshenkov, M.V.; Tcherdyntsev, V.V.; Ergin, K.S.; Shchetinin, I.V. Ultra-high molecular weight polyethylene reinforced with multi-walled carbon nanotubes: Fabrication method and properties. *J. Alloys Compd.* **2012**, *536*, 538–540. [CrossRef]
84. Fonseca, M.A.; Subramani, K.; Oliveira, M.S.A.; Simões, J.A. Enhanced UHMWPE reinforced with mwcnt through mechanical ball-milling. *Defect Diffus. Forum* **2011**, *312–315*, 1238–1243. [CrossRef]
85. Lee, J.K.; Rhee, K.Y.; Lee, J.H. Wear properties of 3-aminopropyltriethoxysilanefunctionalized carbon nanotubes reinforced ultra high molecular weight polyethylene nanocomposites. *Polymer Eng. Sci.* **2010**, *50*, 1433–1439. [CrossRef]
86. Liu, Y.; Sinha, S.K. Wear performances and wear mechanism study of bulk UHMWPE composites with nacre and cnt fillers and pfpe overcoat. *Wear* **2013**, *300*, 44–54. [CrossRef]
87. Martínez-Morlanes, M.J.; Castell, P.; Alonso, P.J.; Martinez, M.T.; Puértolas, J.A. Multi-walled carbon nanotubes acting as free radical scavengers in gamma-irradiated ultrahigh molecular weight polyethylene composites. *Carbon* **2012**, *50*, 2442–2452. [CrossRef]
88. Balani, K.; Anderson, R.; Laha, T.; Andara, M.; Tercero, J.; Crumpler, E.; Agarwal, A. Plasma-sprayed carbon nanotube reinforced hydroxyapatite coatings and their interaction with human osteoblasts *in vitro*. *Biomaterials* **2007**, *28*, 618–624. [CrossRef] [PubMed]

89. Wang, Q.; Zhang, D.; Ge, S. Biotribological behavior of ultra-high molecular weight polyethylene composites containing ti ina hip joint simulator. *Proc. Inst. Mech. Eng. Part J.* **2007**, *221*, 307–313. [CrossRef]
90. Ge, S.; Wang, S.; Huang, X. Increasing the wear resistance of UHMWPE acetabular cups by adding natural biocompatible particles. *Wear* **2009**, *267*, 770–776. [CrossRef]
91. Xie, X.L.; Tang, C.Y.; Chan, K.Y.Y.; Wu, X.C.; Tsui, C.P.; Cheung, C.Y. Wear performance of ultrahigh molecular weight polyethylene/quartz composites. *Biomaterials* **2003**, *24*, 1889–1896. [CrossRef] [PubMed]
92. Liu, Y.; Sinha, S.K. Wear performances of uhmwpe composites with nacre and cnts, and pfpe coatings for bio-medical applications. *Wear* **2013**, *300*, 44–54. [CrossRef]
93. Habibovic, P.; van der Valk, C.M.; van Blitterswijk, C.A.; de Groot, K.; Meijer, G. Influence of octacalcium phosphate coating on osteoinductive properties of biomaterials. *J. Mater. Sci. Mater. Med.* **2004**, *15*, 373–380. [CrossRef] [PubMed]
94. Calderon, V.S.; Sánchez-López, J.C.; Cavaleiro, A.; Carvalho, S. Biotribological behavior of Ag-Zrcxn1-x coatings against uhmwpe for joint prostheses devices. *J. Mech. Behav. Biomed. Mater.* **2015**, *41*, 83–91. [CrossRef] [PubMed]
95. Firouzi, D.; Youssef, A.; Amer, M.; Srouji, R.; Amleh, A.; Foucher, D.A.; Bougherara, H. A new technique to improve the mechanical and biological performance of ultra high molecular weight polyethylene using a nylon coating. *J. Mech. Behav. Biomed. Mater.* **2014**, *32*, 198–209. [CrossRef] [PubMed]
96. Ruan, F.; Bao, L. Mechanical enhancement of uhmwpe fibers by coating with carbon nanoparticles. *Fibers Polymer* **2014**, *15*, 723–728. [CrossRef]
97. Zhang, Y.-C.; He, J.-X.; Wu, H.-Y.; Qiu, Y.-P. Surface characterization of helium plasma treated nano-sio2 sol-gel coated uhmwpe filaments by contact angle experiments and atr-ftir. *J. Fiber Bioeng. Inf.* **2010**, *3*, 50–54. [CrossRef]
98. Chu, P.K.; Chen, J.Y.; Wang, L.P.; Huang, N. Plasma-surface modification of biomaterials. *Mater. Sci. Eng.* **2002**, *36*, 143–206. [CrossRef]
99. Puértolas, J.A.; Martínez-Nogués, V.; Martínez-Morlanes, M.J.; Mariscal, M.D.; Medel, F.J.; López-Santos, C.; Yubero, F. Improved wear performance of ultra high molecular weight polyethylene coated with hydrogenated diamond like carbon. *Wear* **2010**, *269*, 458–465. [CrossRef]
100. He, W.; Gonsalves, K.E.; Halberstadt, C.R. Micro/nanomachining and fabrication of materials for biomedical applications. In *Biomedical Nanostructures*; John Wiley & Sons, Inc.: Hoboken, NJ, USA, 2007; pp. 25–47.
101. Dong, H.; Bell, T. State-of-the-art overview: Ion beam surface modification of polymers towards improving tribological properties. *Surf. Coat. Technol.* **1999**, *111*, 29–40. [CrossRef]
102. Boampong, D.K.; Green, S.M.; Unsworth, A. N+ ion implantation of ti6al4v alloy and UHMWPE for total joint replacement application. *J. Appl. Biomater. Biomech.* **2003**, *1*, 164–171. [PubMed]
103. Qiu, Z.-Y.; Chen, C.; Wang, X.-M.; Lee, I.-S. Advances in the surface modification techniques of bone-related implants for last 10 years. *Regen. Biomater.* **2014**, *1*, 67–79. [CrossRef]
104. Ensinger, W.; Höchbauer, T.; Rauschenbach, B. Treatment uniformity of plasma immersion ion implantation studied with three-dimensional model systems. *Surf. Coat. Technol.* **1998**, *103–104*, 218–221. [CrossRef]
105. Powles, R.C.; McKenzie, D.R.; Fujisawa, N.; McCulloch, D.G. Production of amorphous carbon by plasma immersion ion implantation of polymers. *Diam. Relat. Mater.* **2005**, *14*, 1577–1582. [CrossRef]
106. Shi, W.; Li, X.Y.; Dong, H. Improved wear resistance of ultra-high molecular weight polyethylene by plasma immersion ion implantation. *Wear* **2001**, *250*, 544–552. [CrossRef]
107. Zhang, B.; Huang, W.; Wang, J.; Wang, X. Comparison of the effects of surface texture on the surfaces of steel and UHMWPE. *Tribol. Int.* **2013**, *65*, 138–145. [CrossRef]
108. Kustandi, T.S.; Choo, J.H.; Low, H.Y.; Sinha, S.K. Texturing of UHMWPE surface via nil for low friction and wear properties. *J. Phys. D* **2010**, *43*, 015301. [CrossRef]
109. Tan, J.; Saltzman, W.M. Biomaterials with hierarchically defined micro- and nanoscale structure. *Biomaterials* **2004**, *25*, 3593–3601. [CrossRef] [PubMed]
110. Ito, H.; Kaneda, K.; Yuhta, T.; Nishimura, I.; Yasuda, K.; Matsuno, T. Reduction of polyethylene wear by concave dimples on the frictional surface in artificial hip joints. *J. Arthropl.* **2000**, *15*, 332–338. [CrossRef]
111. Kurella, A.; Dahotre, N.B. Review paper: Surface modification for bioimplants: The role of laser surface engineering. *J. Biomater. Appl.* **2005**, *20*, 5–50. [CrossRef] [PubMed]
112. Chyr, A.; Qiu, M.; Speltz, J.; Jacobsen, R.L.; Sanders, A.P.; Raeymaekers, B. A patterned microtexture to reduce friction and increase longevity of prosthetic hip joints. *Wear* **2014**, *315*, 51–57. [CrossRef] [PubMed]

113. Lopez-Cervantes, A.; Dominguez-Lopez, I.; Barceinas-Sanchez, J.D.; Garcia-Garcia, A.L. Effects of surface texturing on the performance of biocompatible UHMWPE as a bearing material during *in vitro* lubricated sliding/rolling motion. *J. Mech. Behav. Biomed. Mater.* **2013**, *20*, 45–53. [CrossRef] [PubMed]
114. Sagbas, B.; Durakbasa, M.N. Effect of surface patterning on frictional heating of vitamin E blended UHMWPE. *Wear* **2013**, *303*, 313–320. [CrossRef]
115. Wang, X.; Wang, J.; Zhang, B.; Huang, W. Design principles for the area density of dimple patterns. *Proc. Inst. Mech. Eng. Part J.* **2014**. [CrossRef]
116. Kelly, N.H.; Fu, R.H.; Wright, T.M.; Padgett, D.E. Wear damage in mobile-bearing TKA is as severe as that in fixed-bearing TKA. *Clin. Orthop. Relat. Res.* **2011**, *469*, 123–130. [CrossRef] [PubMed]

Article

Evaluation of Two Total Hip Bearing Materials for Resistance to Wear Using a Hip Simulator

Kenneth R. St. John

Department of Biomedical Materials Science, University of Mississippi Medical Center, Jackson, MS 39216, USA; kstjohn@umc.edu; Tel.: +1-601-984-6170; Fax: +1-601-984-6087

Academic Editors: Amir Kamali and J. Philippe Kretzer
Received: 11 February 2015; Accepted: 27 May 2015; Published: 3 June 2015

Abstract: Electron beam crosslinked ultra high molecular weight polyethylene (UHMWPE) 32 mm cups with cobalt alloy femoral heads were compared with gamma-irradiation sterilized 26 mm cups and zirconia ceramic heads in a hip wear simulator. The testing was performed for a total of ten million cycles with frequent stops for cleaning and measurement of mass losses due to wear. The results showed that the ceramic on UHMWPE bearing design exhibited higher early wear than the metal on highly crosslinked samples. Once a steady state wear rate was reached, the wear rates of the two types of hip bearing systems were similar with the ceramic on UHMPWE samples continuing to show a slightly higher rate of wear than the highly crosslinked samples. The wear rates of each of the tested systems appear to be consistent with the expectations for low rates of wear in improved hip replacement systems.

Keywords: UHMWPE; wear simulation; alternative bearings; hip prosthesis

---

1. Introduction

The wear of ultra high molecular weight polyethylene acetabular cups after total hip replacement surgery has been an issue of concern and study for many years [1–6]. After the discovery of high rates of wear of some polyethylene components in some patients, efforts began to try to improve the wear rates through design changes, polymer modification [7–11], or materials substitution [12–14]. Changes included the use of ceramic femoral head components to reduce wear, modification of the processing, packaging, and sterilization parameters of the acetabular cups, varying the molecular weight of the polyethylene or changing the additives, and modification of the polyethylene material itself.

In attempts to improve the wear resistance of the femoral head and acetabular cup bearing couple, attention has focused on both sides of the interface. Changes have been made to reduce wear of titanium alloy heads by modifying the surface [12,13] so that it would not be necessary to utilize cobalt heads on titanium alloy femoral stems. It has been suggested that the nature of the femoral head might affect the wear of the polyethylene so a change to ceramic heads [15–24] was tested and has been instituted in many marketed devices. The use of ceramic-on-ceramic, in which both component materials are replaced by ceramic has also been tested and utilized in surgery [25–28] for several years. Metal-on-metal articulations initially had been used in the 1960s and 1970s [29,30] and have recently become attractive again as a possible solution to wear problems [22,31–38].

This preliminary screening study compares the wear properties of an electron beam crosslinked ultra high molecular weight polyethylene (UHMWPE) when bearing against a 32 mm cobalt/chromium alloy femoral head with the properties of a gamma-irradiation-sterilized UHMWPE utilizing a 26 mm ceramic head. Previous clinical studies have shown that head size had an effect on wear rates of gamma-irradiation-sterilized UHMWPE [39,40] with larger heads exhibiting greater linear and volumetric wear rates. However, larger femoral heads have been advocated to reduce dislocation [41,42] and improve the range of motion and stability [41,43,44] of total hip replacements. Research has shown that the crosslinking of UHMWPE will reduce the wear rates in both laboratory testing [45–49] and in clinical experience [46,50–53]. It has been suggested that the detrimental effect of head size on wear rates may not exist for highly crosslinked UHMWPE [54,55], which may allow for the use of larger femoral heads to achieve the potential benefits. Finite element analysis modelling the predicted stresses in conventional and highly crosslinked polyethylene and Fuji film pressure analysis suggested that the stresses in highly crosslinked UHMWPE with large head sizes might be less than for 28 mm conventional polyethylene [56]. A clinical evaluation [57], with the crosslinked UHMWPE that is a part of this study, suggests that larger head sizes do produce increased wear volume over smaller heads, meaning that the use of crosslinked polyethylene, while it may reduce the wear rates over non-crosslinked material, still may suffer from the effect of increasing head size. Another clinical study showed that a wear increase due to increased head size did not appear to be significant [58].

The use of ceramic femoral heads has been suggested as a way to reduce the wear of the UHMWPE acetabular components [59,60]. Laboratory studies [15,23,61] and clinical studies [15,20,22,62] showing a wear rate improvement conflict with other clinical reports [63–65] suggesting no apparent difference. Recently, a matched pair analysis was reported comparing 28 mm ceramic heads with the same size metal heads [66], bearing on conventional UHMWPE and found a significant reduction in wear rate at an average of 17 years post implantation, when ceramic heads were used.

The sponsor of this study had requested the comparison of two different designs of hip replacements that he was considering using in his practice. The designs tested were the two specific designs for which he wished to compare the wear properties. The larger head and crosslinked polyethylene were available for the metal on polyethylene device and might offer advantages in reducing hip dislocation. One design utilized a large head bearing against a polyethylene that was claimed to have improved wear properties due to a crosslinking process. The other design contained a more conventional smaller head and no specific processing to enhance wear properties but the head was manufactured from a ceramic material, presumably conferring improved wear characteristics to the bearing pair.

In addition to the differences in bearing materials of the two different device designs which were tested, the head sizes were different as well. The head size for the ceramic heads was 26 mm, while that for the cobalt alloy heads was 32 mm. For one of the devices, research had shown that there might be no significant difference in the clinical wear rate for the Longevity® device when 26 mm heads were compared to 32 mm heads, although there was a trend towards a higher wear rate for the larger head [58]. For the ceramic on polyethylene, no specific study has compared the effects of head size on wear rates, although a study has been reported in which 22.25, 26, and 28 mm alumina ceramic heads from the same manufacturer were compared in a hip simulator study [40]. This study showed increases in the wear rate when head size was increased and also reviewed published literature showing that other researchers had reported studies in which larger heads had lower wear, smaller heads showed lower wear, or the wear rates were equivalent.

2. Materials and Methods

The devices tested were either 32 mm Co/Cr femoral heads bearing upon Zimmer® Longevity® polyethylene cups [67] or 26 mm Kyocera® Zirconia heads bearing upon Kyocera polyethylene cups. The 26 mm cups were manufactured from GUR 1020 resin and sterilized by gamma irradiation (4.5 MRad). The 32 mm cups were prepared and crosslinked according to a proprietary process said to reduce the wear rate between 88% and 98% over non-crosslinked polyethylene. ("Longevity™ Crosslinked Polyethylene", Zimmer Informational Brochure).

Three of each type of cup was tested in wear and one of each type was subjected to loading, but not movement. The polyethylene cups that were subjected loading only served as soak controls in an effort to correct for the effects of fluid uptake on weight changes of the cups. The cups were mounted in a heat-softenable polyurethane resin using extra cups to avoid subjecting the test cups to the 80 °C temperatures required to cure the polyurethane.

Testing was conducted on an MTS® eight-station hip simulator, which has been customized to add load, torque, and displacement transducers on all eight stations. This simulator is of the type frequently called "Orbital Bearing", which is the subject of ISO standard ISO 14242-3 [68]. The wear path described is the same for every loading cycle and attempts to mimic part of the normal motion of a hip in use. The devices were mounted with the head and fixture on the bottom and the cup on top (anatomic position). A Paul loading curve with a maximum load of 3000 N was used and loading synchronized with the rotation of the devices so that the same load was placed upon the same area of the cup during each cycle. The lubricant for each station was 50% bovine calf serum (Hyclone) diluted with deionized water and to which was added 20 mM Disodium EDTA Dihydrate. The EDTA was added to the deionized water and thoroughly mixed. The water/EDTA solution was filtered through 0.2 μm filter disks before being mixed with the serum.

While it was not possible to measure interface friction in these samples directly, torsional load cells were mounted on each testing station and the maximum and minimum torque measured for each sample was recorded over a total of approximately 10 cycles, periodically during testing. The difference between the highest and lowest torque was calculated and expressed as the total torsional excursion during the loading pattern applied.

Testing was conducted to a total of ten million cycles at a rate of one cycle per second (approximately 86,000 cycles per day). Measurement of weight loss during the study was conducted at every 500,000 cycles. The test was stopped and the serum removed and stored. At each cleaning interval, the cups were dismounted from the fixturing, thoroughly cleaned in accordance with internal laboratory protocols and ASTM F1714 [69], dried, and then weighed three times in a round-robin fashion to determine the weight changes which had occurred. Weight loss values were corrected for fluid uptake based upon changes in weight of the loaded soak control cups. A balance readable to 0.01 mg was used for all weighings. This method of characterizing losses due to wear is very reproducible and is consistent with the international standards [69,70].

Each polyethylene cup was inspected on a laser scanning confocal microscope (Leica SP-2, Leica Microsystems, Exton, PA, USA) and an image collected using the 10 × objective, giving a field of view of 1.5 mm × 1.5 mm (2.25 mm^2). The use of the confocal microscope allowed the viewing of the non-conductive polyethylene without the charging problems that would be experienced in an electron microscope. The surface morphology of the worn cups was compared to that for the as-manufactured cups that had been subjected only to loading without oscillating movement.

3. Results

Weight losses due to wear began to occur during the first 500,000 cycles in the case of the ceramic on polyethylene samples. There was very little detectable weight loss for the crosslinked cups for the first 1,000,000 cycles but, after 1,000,000 cycles, there were losses at an average rate only slightly less than that for the ceramic on polyethylene and that average was the result of one crosslinked cup experiencing much less wear than the other two. The weight loss results over ten million cycles are shown in Figure 1 for each tested cup. In Figures 2 and 3, the wear results are averaged for each type of bearing pair. The average wear rates for each type of sample are given in Table 1. In the last column, it can be seen that the average wear rate (mg/mc) for the last 8,000,000 cycles was 3.55 ± 0.55 for the ceramic on polyethylene and 3.20 ± 0.88 for the cross-linked materials. Throughout the study the average wear rate for the crosslinked polyethylene remained slightly less than that for the ceramic on polyethylene but two out of the three crosslinked cups had slightly higher wear rates than the ceramic/polyethylene bearings in the last eight million cycles. The third crosslinked cup had a much lower wear rate than the other two samples but no differences were noted in the appearance of the cup and head or the appearance of the testing lubricant. One of the ceramic-on-polyethylene had a higher total wear than the other two but the difference was not as large as that for the crosslinked cups. In these samples as well, no differences were observed that would explain the difference.

All of the tested devices were provided from inventory ready for implantation and one would assume that all devices met the manufacturer's specifications but variations in head and cup diameters could have played a role in the observed results. The diameter data is not available for the individual samples, so this comparison was not performed.

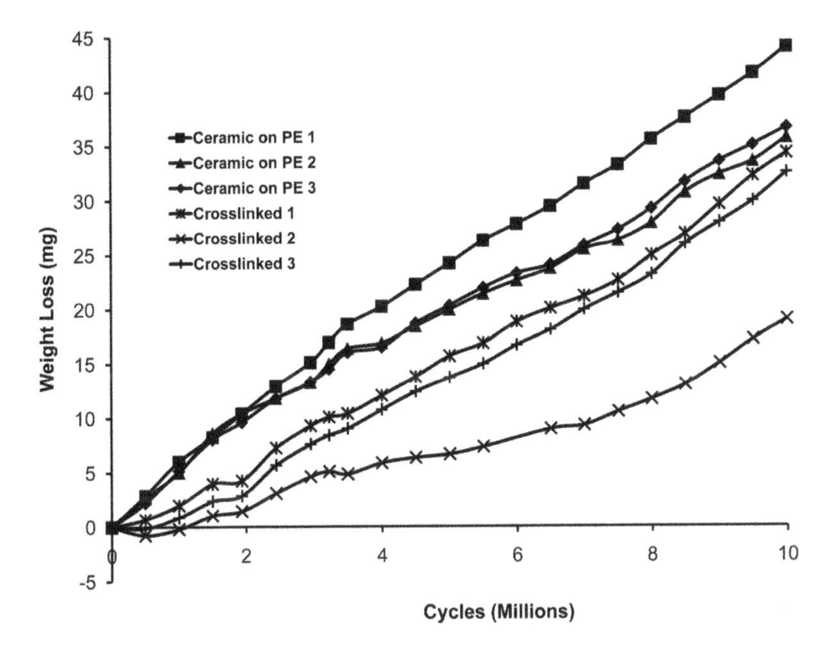

Figure 1. Individual sample wear data (load soak corrected) for each of the six tested samples.

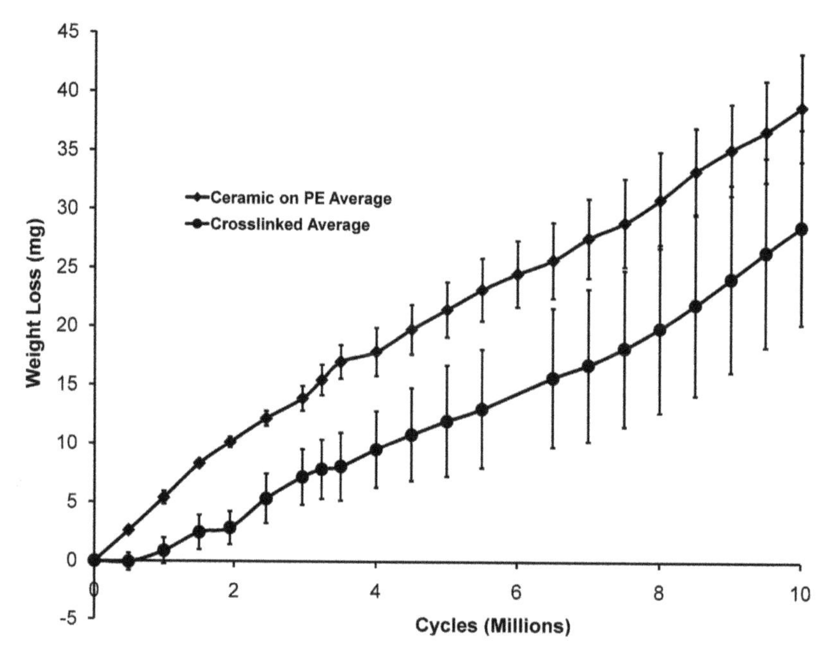

Figure 2. Averaged wear data for each of the two sample types.

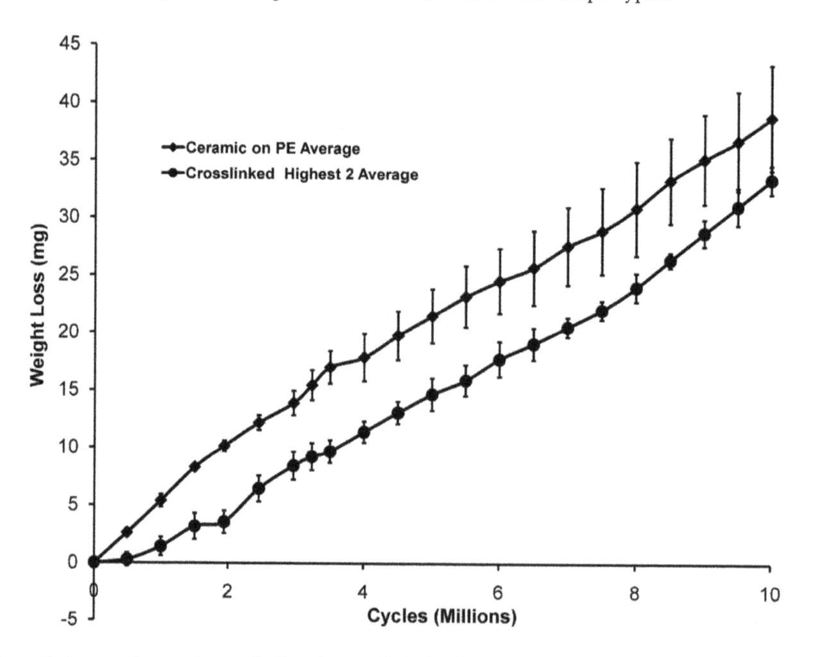

Figure 3. Averaged wear data excluding the crosslinked polyethylene sample with the lowest wear rate.

Table 1. Wear rates at different points in the testing (milligrams per million cycles).

Sample Type	First 1,000,000	1,000,000 to 2,000,000	2,000,000 to 5,000,000	5,000,000 to 10,000,000	2,000,000 to 10,000,000
Ceramic on Polyethylene	5.40 ± 0.54	5.04 ± 0.59	3.71 ± 0.72	3.46 ± 0.44	3.55 ± 0.55
Crosslinked	0.89 ± 1.10	2.08 ± 0.33	2.99 ± 1.12	3.32 ± 0.73	3.20 ± 0.88
Crosslinked (2 highest)	1.43	2.26	3.63	3.74	3.70

During each loading cycle, the torque measured varied from clockwise to counterclockwise dependent upon the magnitude and direction of the loads being applied at that moment. Measurements of the torque excursion during testing showed that the friction at the bearing surfaces of the ceramic on polyethylene bearings was initially much higher than that for the metal on crosslinked polymer. This is illustrated in Figure 4. As the testing progressed, this situation was reversed and, as the wear rate of the crosslinked cups increased, the friction increased as well. From 3,000,000 cycles to 5,000,000 cycles, the torques for the two types of bearings were essentially equivalent. After 5,000,000 cycles, the average torques for the crosslinked cups were higher than for the ceramic on polyethylene bearings at all except one time point.

Figures 5–9 document the appearance of five of the eight polyethylene cups after 10,000,000 cycles of testing. Figures 6 and 9 show the surface morphology of the acetabular cups that were used a soak controls for the testing, showing the machining marks that were present in all cups before any wear occurred. Manufactured polyethylene cups have residual machining marks as a result of the manufacturing process and, in most cases; the machining marks have completely disappeared by 10,000,000 cycles (Figures 5 and 7). One of the tested cups (Figure 8) appears to have retained a portion of the machining marks after 10,000,000 cycles.

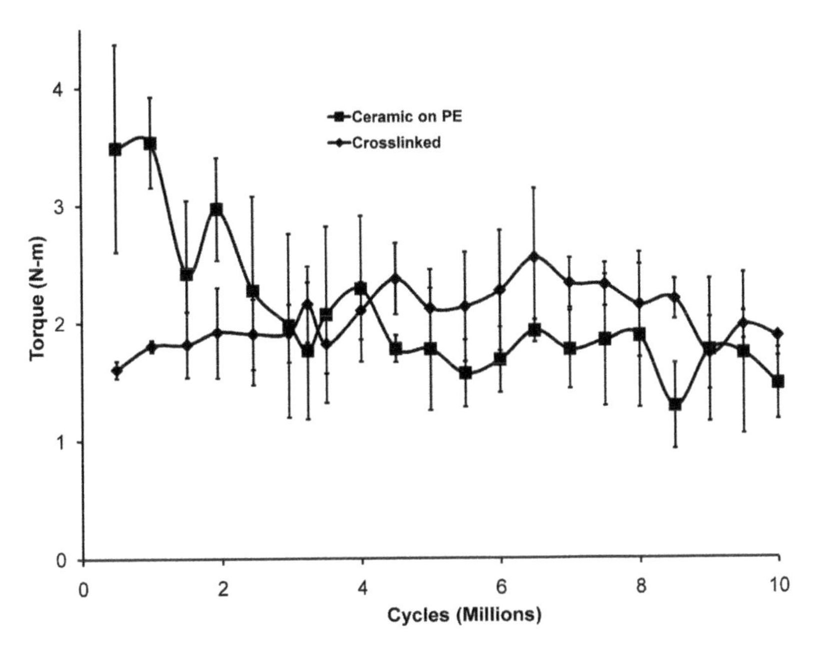

Figure 4. Measured torque values averaged for each of the two sample types.

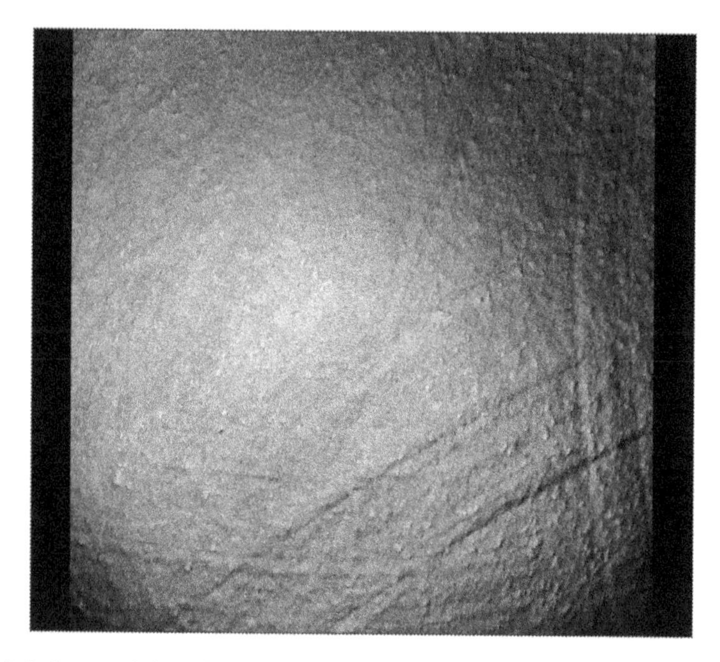

Figure 5. Surface morphology of an acetabular cup that has been tested with a zirconia femoral head.

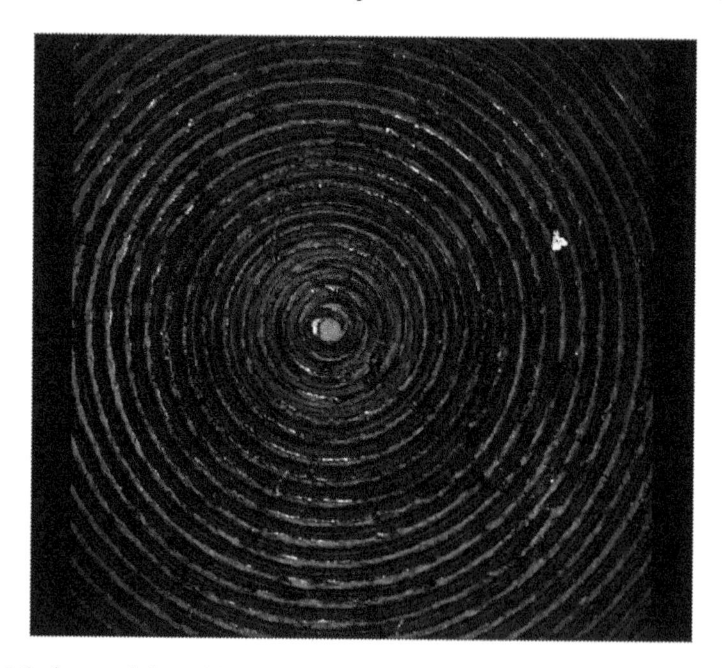

Figure 6. Surface morphology of the same type of cup as in Figure 5 in the as-manufactured condition.

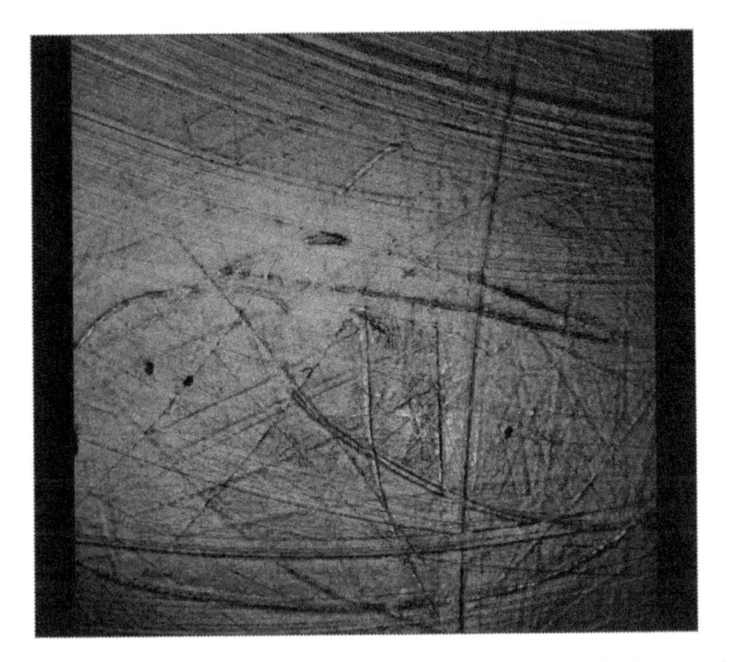

Figure 7. Surface morphology of a crosslinked polyethylene acetabular cup that has been tested with a cobalt alloy femoral head.

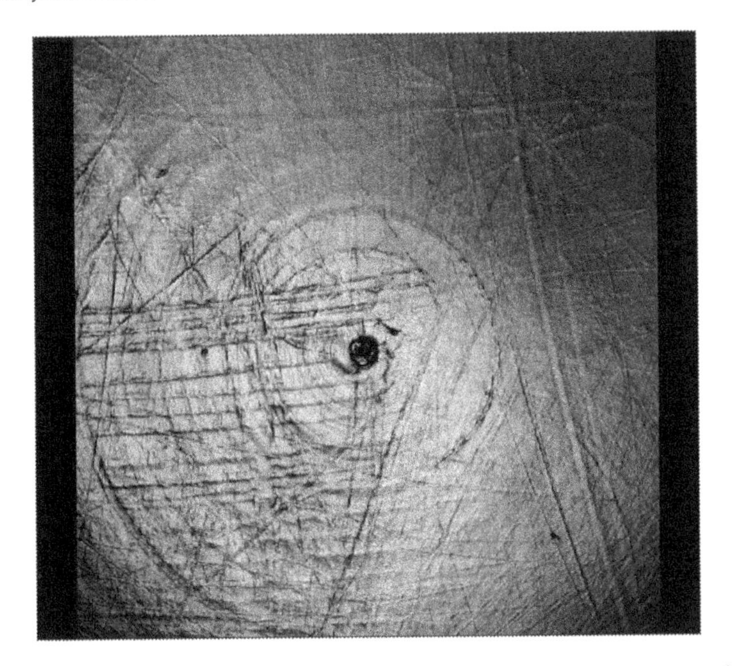

Figure 8. Surface morphology of a crosslinked polyethylene acetabular cup that has been tested with a cobalt alloy femoral head, showing residual machining marks after 10 million wear cycles.

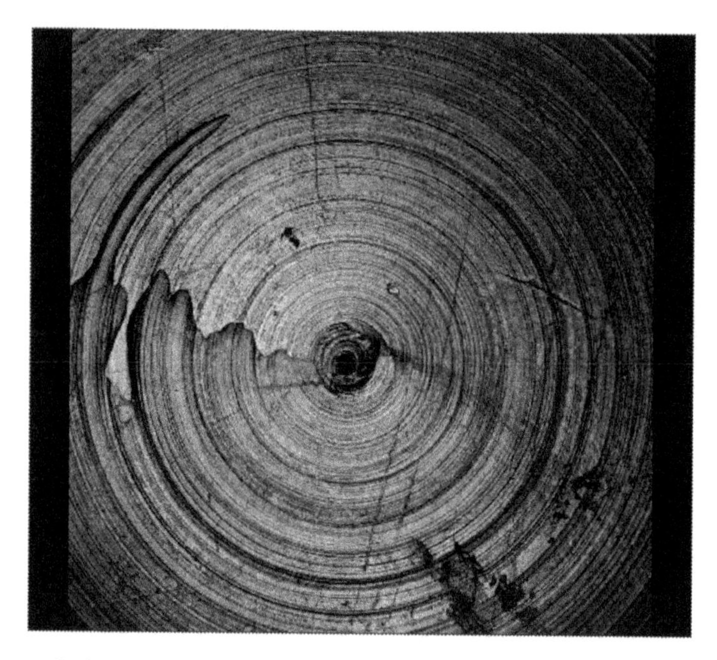

Figure 9. Surface morphology of the same type of cup as in Figures 7 and 8 in the as-manufactured condition.

The microphotography of the tester cups did not show any observable differences in surface morphology that would suggest an explanation for one of the samples with a ceramic head having higher wear results and one of the crosslinked samples having a much lower wear result.

4. Discussion

The total wear seen with the Kyocera ceramic head bearing on polyethylene was greater than that for Zimmer metal head on crosslinked polyethylene but, after about 2,000,000 cycles, the average wear rates for the two types of bearings were within one standard deviation of each other and the average wear rate of the two crosslinked cups with the highest wear rates was equivalent to the rate using ceramic heads.

The torques transmitted to the torque cells due to the movement at the bearing surfaces started out much higher for the ceramic-on-polyethylene bearings but decreased to a level slightly less than that for metal-on-crosslinked polyethylene by about 4,500,000 cycles and then remained relatively constant for the remainder of the testing period. The torques for the metal-on-crosslinked polyethylene were nearly constant throughout. This may be due to the differences in the initial surface morphology after machining or the differences in surface roughness of the head after manufacture. Over time, the two sides of the bearing couple apparently become accommodated to each other and it appears that the wear mechanism changes, perhaps to mixed mode fluid film lubrication. While this measurement is not a direct measurement of the frictional forces, the measurements are correlated with the frictional forces being experienced since the magnitude of the torque will be proportional with the force required to move the bearing surfaces past each other, which is the frictional force. As a result of the fact that the head sizes of the two devices are different, drawing conclusions about the relative friction between the two designs would be inappropriate but the conclusion can be drawn that the initial friction in the 26 mm bearing couples was much greater early in testing than after about 3 million cycles, while the friction in the 32 mm bearings was relatively constant over the entire experiment. No attempt has been made to determine the actual friction or the effect of head size on friction, although it would seem that the same frictional force at a greater distance from the center of rotation (16 mm *vs.* 13 mm) would yield a higher torque, which might explain the relationship between the measured torques after about 3 million cycles. Current activity in ASTM subcommittee F04.22 on Arthroplasty has included efforts to describe the relationship between measured torques and frictional forces at the bearing surface when devices are tested in hip simulators.

Researchers have reported [54,71] that the rate of wear of the electron beam crosslinked polyethylene after hip simulator testing for the equivalent of 12 years was not measurable under the conditions of their testing. In this study, two of the highly crosslinked cups did not exhibit measurable wear for the first 500,000 cycles (Figure 1) and the third almost no wear. At 1 million cycles, one was still not showing measurable wear, although all three began to show measurable wear by 1,500,000 cycles. Once the highly crosslinked cups began to show wear, the steady state wear rates for two of the samples were very similar to those for the ceramic on polyethylene tested samples. The third crosslinked cup showed much lower wear rates and still exhibited some residual machining marks after 10,000,000 cycles (Figure 8). The surface morphology of the samples after testing is very similar to that which has been previously reported [71]. Clinical results with these devices confirmed significantly reduced wear rates as compared to devices that had not received the crosslinking treatment [52,72].

The author is not aware of any other studies comparing these two designs in a single laboratory or clinical investigation. A study published in 1995 comparing the wear of highly irradiated UHMWPE with metallic heads to normally sterilized polyethylene with ceramic heads showed very similar but slightly higher wear rates for the ceramic head samples [51], which is consistent with the results of this study.

5. Conclusions

The results of hip simulator wear testing of highly crosslinked UHMWPE 32 mm cups with metallic femoral heads as compared to gamma-irradiation sterilized 26 mm cups with zirconia ceramic heads suggests that the ceramic on polyethylene devices begin to wear sooner and had higher total wear results. After the initial wearing in period, it appears that the steady state wear rates may be very similar. The wear rates of each of the tested bearing combinations is well within the rate of low wear rates that have been reported for improved hip bearing designs.

Based upon these preliminary results, either design is likely to yield acceptable rates of bearing wear when used in total joint replacement surgery.

Acknowledgments: The author would like to thank Masaaki Matsubara, for financial support for this research and Japan Medical Materials and Zimmer Japan for donation of test specimens and fixturing.

Conflicts of Interest: The author received funding for this research from a non-commercial source (Masaaki Matsubara) and received nothing of value from any manufacturer or commercial source other than the donation of the devices tested in this study.

References

1. Cates, H.E. Polyethylene wear in cemented metal-backed actabular cups. *J. Bone Joint Surg. Br.* **1993**, *75*, 249–253. [PubMed]
2. Galante, J.O.; Rostoker, W. Wear in total hip prostheses. *ACTA Orthop. Suppl.* **1973**, *145*, 1–46.
3. Jasty, M.; Goetz, D.D.; Bragdon, C.R.; Lee, K.R.; Hanson, A.E.; Elder, J.R.; Harris, W.H. Wear of polyethylene acetabular components in total hip arthroplasty. *J. Bone Joint Surg. Am.* **1997**, *79*, 349–358. [PubMed]
4. Kabo, J.M. *In Vivo* Wear of polyethylene acetabular components. *J. Bone Joint Surg. Br.* **1993**, *75*, 254–258. [PubMed]
5. Lewis, G. Polyethylene wear in total hip and knee arthroplasties. *J. Biomed. Mater. Res.* **1997**, *38*, 55–75. [CrossRef]

6. Schmalzried, T.P.; Callaghan, J.J. Wear in total hip and knee replacements. *J. Bone Joint Surg. Am.* **1999**, *81*, 115–136. [PubMed]
7. Edidin, A.A.; Herr, M.P.; Villarraga, M.L.; Muth, J.; Yau, S.S.; Kurtz, S.M. Accelerated aging studies of UHMWPE. I. Effect of resin, processing, and radiation environment on resistance to mechanical degradation. *J. Biomed. Mater. Res.* **2002**, *61*, 312–322. [CrossRef] [PubMed]
8. Elfick, A.P.D. The effect of socket design, materials and liner thickness on the wear of the porous coated anatomic total hip replacement. *Proc. Inst. Mech. Eng.* **2001**, *215*, 447–457. [CrossRef]
9. Liu, C.Z.; Wu, J.Q.; Li, J.Q.; Ren, L.Q.; Tong, J.; Arnell, A.D. Tribological behaviours of PA/UHMWPE blend under dry and lubricating condition. *Wear* **2006**, *260*, 109–115. [CrossRef]
10. McKellop, H.; Shen, F.W.; Lu, B.; Campbell, P.; Salovey, R. Effect of sterilization method and other modifications on the wear resistance of acetabular cups made of ultra-high molecular weight polyethylene. A hip-simulator study. *J. Bone Joint Surg. Am.* **2000**, *82*, 1708–1725. [PubMed]
11. McKellop, H.A.; Shen, F.-W.; Campbell, P.; Ota, T. Effect of molecular weight, calcium stearate, and sterilization methods on the wear of ultra high molecular weight polyethylene acetabular cups in a hip joint simulator. *J. Orthop. Res.* **1999**, *17*, 329–339. [CrossRef] [PubMed]
12. Allen, C.; Bloyce, A.; Bell, T. Sliding wear behaviour of ion implanted ultra high molecular weight polyethylene against a surface modified titanium alloy Ti-6Al-4V. *Trib. Int.* **1996**, *29*, 527–534. [CrossRef]
13. Rostland, T.; Albrektsson, B.; Albrektsson, T.; McKellop, H. Wear of ion-implanted pure titanium against UHMWPE. *Biomaterials* **1989**, *10*, 176–181. [CrossRef]
14. Scholes, S.C. Compliant layer acetabular cups: Friction testing of a range of materials and designs for a new generation of prosthesis that mimics the natural joint. *Proc. Inst. Mech. Eng.* **2006**, *220*, 583–596. [CrossRef]
15. Clarke, I.C.; Gustafson, A. Clinical and hip simulator comparisons of ceremic-on-polyethylene and metal-on-polethylene wear. *Clin. Orthop.* **2000**, *379*, 34–40. [CrossRef] [PubMed]
16. Galvin, A.; Brockett, C.; Williams, S.; Hatto, P.; Burton, A.; Isaac, G.; Stone, M.; Ingham, E.; Fisher, J. Comparison of wear of ultra-high molecular weight polyethylene acetabular cups against surface-engineered femoral heads. *Proc. Inst. Mech. Eng.* **2008**, *222*, 1073–1080. [CrossRef]
17. Goldsmith, A.A.; Dowson, D. A multi-station hip joint simulator study of the performance of 22 mm diameter zirconia-ultra-high molecular weight polyethylene total replacement hip joints. *Proc. Inst. Mech. Eng.* **1999**, *213*, 77–90. [CrossRef]
18. Hernigou, P.; Bahrami, T. Zirconia and alumina ceramics in comparison with stainless-steel heads: Polyethylene wear after a minimum ten-year follow-up. *J. Bone Joint Surg. Br.* **2003**, *85*, 504–509. [CrossRef] [PubMed]
19. Kawanabe, K.; Tanaka, K.; Tamura, J.; Shimizu, M.; Onishi, E.; Iida, H.; Nakamura, T. Effect of alumina femoral head on clinical results in cemented total hip arthroplasty: Old *vs.* current alumina. *J. Orthop. Sci.* **2005**, *10*, 378–384. [CrossRef] [PubMed]
20. Kim, Y.H. Comparison of polyethylene wear associated with cobalt-chromium and zirconia heads after total hip replacement. A prospective, randomized study. *J. Bone Joint Surg. Am.* **2005**, *87*, 1769–1776. [CrossRef] [PubMed]
21. Saikko, V.; Ahlroos, T.; Calonius, O.; Keranen, J. Wear simulation of total hip prostheses with polyethylene against CoCr, alumina and diamond-like carbon. *Biomaterials* **2001**, *22*, 1507–1514. [CrossRef]
22. Semlitsch, M.; Willert, H.G. Clinical wear behavior of ultra-high molecular weight polyethylene cups paired with metal and ceramic ball heads in comparison to metal-on-metal pairings of hip joint replacements. *Proc. Inst. Mech. Eng.* **1997**, *211*, 73–88. [CrossRef]
23. Smith, S.L.; Unsworth, A. A comparison between gravimetric and volumetric techniques of wear measurement of UHMWPE acetabular cups against zirconia and coblat-chromium-molybdenum femoral heads in a hip simulator. *Proc. Inst. Mech. Eng.* **1999**, *213*, 475–483. [CrossRef]
24. Wroblewski, B.M.; Siney, P.D.; Fleming, P.A. Low-friction arthroplasty of the hip using alumina ceramic and cross-linked polyethylene. *J. Bone Joint Surg. Br.* **1999**, *81*, 54–55. [CrossRef] [PubMed]
25. Affatato, S.; Ferrari, G.; Chevalier, J.; Ruggeri, O.; Toni, A. Surface characterization and debris analysis of ceramic pairings after ten million cycles on a hip joint simulator. *Proc. Inst. Mech. Eng.* **2002**, *216*, 419–424. [CrossRef]

26. Shishido, T.; Clarke, I.C.; Williams, P.; Boehler, M.; Asano, T.; Shoji, H.; Masaoka, T.; Yamamoto, K.; Imakiire, A. Clinical and simulator wear study of alumina ceramic THR to 17 years and beyond. *J. Biomed. Mater. Res.* **2003**, *67*, 638–647. [CrossRef] [PubMed]
27. Stewart, T.D.; Tipper, J.L.; Insley, G.; Streicher, R.M.; Ingham, E.; Fisher, J. Long-term wear of ceramic matrix composite materials for hip prostheses under severe swing phase microseparation. *J. Biomed. Mater. Res.* **2003**, *66*, 567–573. [CrossRef] [PubMed]
28. Yoon, T.R.; Rowe, S.M.; Jung, S.T.; Seon, K.J.; Maloney, W.J. Osteolysis in association with a total hip arthroplasty with ceramic bearing surfaces. *J. Bone Joint Surg. Am.* **1998**, *80*, 1459–1468. [PubMed]
29. Muller, M.E. The benefits of metal-on-metal total hip replacements. *Clin. Orthop.* **1995**, *311*, 54–59. [PubMed]
30. Walker, P.S.; Erkman, M.J. Metal-on-metal lubrication in artificial human joints. *Wear* **1972**, *21*, 377–392. [CrossRef]
31. Chan, F.W.; Bobyn, J.D.; Medley, J.B.; Krygier, J.J.; Tanzer, M. The Otto Aufranc Award. Wear and lubrication of metal-on-metal hip implants. *Clin. Orthop. Relat. Res.* **1999**, *369*, 10–24. [CrossRef] [PubMed]
32. Dorr, L.D.; Wan, Z.; Longjohn, D.B.; Dubois, B.; Murken, R. Total hip arthroplasty with use of the Metasul metal-on-metal articulation. Four to seven-year results. *J. Bone Joint Surg. Am.* **2000**, *82*, 789–798. [PubMed]
33. Fisher, J.; Hu, X.Q.; Stewart, T.D.; Williams, S.; Tipper, J.L.; Ingham, E.; Stone, M.H.; Davies, C.; Hatto, P.; Bolton, J.; *et al.* Wear of surface engineered metal-on-metal hip prostheses. *J. Mater. Sci. Mater. Med.* **2004**, *15*, 225–235. [CrossRef] [PubMed]
34. Fisher, J.; Hu, X.Q.; Tipper, J.L.; Stewart, T.D.; Williams, S.; Stone, M.H.; Davies, C.; Hatto, P.; Bolton, J.; Riley, M.; *et al.* An *in vitro* study of the reduction in wear of metal-on-metal hip prostheses using surface-engineered femoral heads. *Proc. Inst. Mech. Eng.* **2002**, *216*, 219–230. [CrossRef]
35. Goldsmith, A.A.; Dowson, D.; Isaac, G.H.; Lancaster, J.G. A comparative joint simulator study of the wear of metal-on-metal and alternative material combinations in hip replacements. *Proc. Inst. Mech. Eng.* **2000**, *214*, 39–47. [CrossRef]
36. Schmalzried, T.P.; Peters, P.C.; Maurer, B.T.; Bragdon, C.R.; Harris, W.H. Long-duration metal-on-metal total hip arthroplasties with low wear of the articulating surfaces. *J. Arthroplast.* **1996**, *11*, 322–331. [CrossRef]
37. Streicher, R.M.; Semlitsch, M.; Schon, R.; Weber, H.; Rieker, C. Metal-on-metal articulation for artificial hip joints: Laboratory study and clinical results. *Proc. Inst. Mech. Eng.* **1996**, *210*, 223–232. [CrossRef]
38. Willert, H.G.; Buchhorn, G.H.; Gobel, D.; Koster, G.; Schaffner, S.; Schenk, R.; Semlitsch, M. Wear behavior and histopathology of classic cemented metal on metal hip endoprostheses. *Clin. Orthop.* **1996**, *329*, S160–S186. [CrossRef] [PubMed]
39. Livermore, J.; Ilstrup, D.; Morrey, B. Effect of femoral head size on wear of the polyethylene acetabular component. *J. Bone Joint Surg. Am.* **1990**, *72*, 518–528. [PubMed]
40. Clarke, I.C.; Gustafson, A.; Jung, H.; Fujisawa, A. Hip-simulator ranking of polyethylene wear: Comparisons between ceramic heads of different sizes. *Acta Orthop. Scand.* **1996**, *67*, 128–132. [CrossRef] [PubMed]
41. Beaule, P.E.; Schmalzreid, T.P.; Udomkiat, P.; Amstutz, H.C. Jumbo femoral head for the treatment of recurrent dislocation following total hip replacement. *J. Bone Joint Surg. Am.* **2002**, *84*, 256–263. [PubMed]
42. Kluess, D.; Martin, H.; Mittelmeier, W.; Schmitz, K.-P.; Bader, R. Influence of femoral head size on impingement, dislocation and stress distribution in total hip replacement. *Med. Eng. Phys.* **2007**, *29*, 465–471. [CrossRef] [PubMed]
43. Burroughs, B.R.; Hallstrom, B.; Golladay, G.J.; Hoeffel, D.; Harris, W.H. Range of motion and stability in total hip arthroplasty with 28-, 32-, 38-, and 22-mm femoral head sizes. *J. Arthroplast.* **2005**, *20*, 11–19. [CrossRef] [PubMed]
44. Matsushita, A.; Nakashima, Y.; Jingushi, S.; Yamamoto, T.; Kuraoka, A.; Iwamoto, Y. Effects of the femoral offset and the head size on the safe range of motion in total hip arthroplasty. *J. Arthroplast.* **2009**, *24*, 646–651. [CrossRef] [PubMed]
45. D'Lima, D.D.; Hermida, J.C.; Chen, P.C.; Colwell, C.W., Jr. Polyethylene cross-linking by two different methods reduces acetabular liner wear in a hip joint wear simulator. *J. Orthop. Res.* **2003**, *21*, 761–766. [CrossRef]
46. Geerdink, C.H.; Grimm, B.; Ramakrishnan, R.; Ronduis, J.; Verberg, A.J.; Tonino, A.J. Crosslinked polyethylene compared to conventional polyethylene in total hip replacement: Pre-clinical evaluation, *in vitro* testing and prospective clinical follow-up study. *Acta Orthop.* **2006**, *77*, 719–725. [CrossRef] [PubMed]

47. McKellop, H.; Shen, F.-W.; Lu, B.; Campbell, P.; Salovey, R. Development of an extremely wear-resistant ultra high molecular weight polyethylene for total hip replacements. *J. Orthop. Res.* **1999**, *17*, 157–167. [CrossRef] [PubMed]
48. Muratoglu, O.K.; O'Connor, D.O.; Bragdon, C.R.; Delaney, J.; Jasty, M.; Harris, W.H.; Merrill, E.; Venugopalan, P. Gradient crosslinking of UHMWPE using irradiation in molten state for total joint arthroplasty. *Biomaterials* **2002**, *23*, 717–724. [CrossRef]
49. Muratoglu, O.K.; Wannomae, K.; Christensen, S.; Rubash, H.E.; Harris, W.H. *Ex vivo* wear of conventional and cross-linked polyethylene acetabular liners. *Clin. Orthop.* **2005**, *438*, 158–164. [CrossRef] [PubMed]
50. Gordon, A.C.; D'Lima, D.D.; Colwell, C.W. Highly cross-linked polyethylene in total hip arthroplasty. *J. Am. Acad. Orthop. Surg.* **2006**, *14*, 511–523. [PubMed]
51. Oonishi, H.; Takayama, Y.; Tsuji, E. The low wear of cross-linked polyethylene socket in total hip prostheses. In *Encyclopedic Handbook of Biomaterials and Bioengineering, Part A: Materials*; Wise, D.L., Trantolo, D.J., Altobelli, D.F., Yaszemski, M.J., Gresser, J.D., Schwartz, E.R., Eds.; Marcel Dekker: New York, NY, USA, 1995; Volume 1, pp. 1853–1868.
52. McCalden, R.W.; MacDonald, S.J.; Rorabeck, C.H.; Bourne, R.B.; Chess, D.G.; Charron, K.D. Wear rate of highly cross-linked polyethylene in total hip arthroplasty. A randomized controlled trial. *J. Bone Joint Surg. Am.* **2009**, *91*, 773–782. [CrossRef] [PubMed]
53. Nikolaou, V.S.; Edwards, M.R.; Bogoch, E.; Schemitsch, E.H.; Waddell, J.P. A prospective randomised controlled trial comparing three alternative bearing surfaces in primary total hip replacement. *J. Bone Joint Surg. Br.* **2012**, *94*, 459–465. [CrossRef] [PubMed]
54. Muratoglu, O.K.; Bragdon, C.R.; O'Connor, D.; Perinchief, R.S.; Estok, D.M.; Jasty, M.; Harris, W.H. Larger diameter femoral heads used in conjunction with a highly cross-linked ultra-high molecular weight polyethylene: A new concept. *J. Arthroplast.* **2001**, *16*, 24–30. [CrossRef]
55. Hermida, J.C.; Bergula, A.; Chen, P.; Colwell Jr, C.W.; D'Lima, D.D. Comparison of the wear rates of twenty-eight and thirty-two millimeter femoral heads on cross-linked polyethylene acetabular cups in a wear simulator. *J. Bone Joint Surg. Am.* **2003**, *85*, 2325–2331. [PubMed]
56. Plank, G.R.; Estok, D.M., 2nd; Muratoglu, O.K.; O'Connor, D.O.; Burroughs, B.R.; Harris, W.H. Contact stress assessment of conventional and highly crosslinked ultra high molecular weight polyethylene acetabular liners with finite element analysis and pressure sensitive film. *J. Biomed. Mater. Res. B Appl. Biomater.* **2007**, *80*, 1–10. [CrossRef] [PubMed]
57. Lachiewicz, P.F.; Heckman, D.S.; Soileau, E.S.; Mangla, J.; Martell, J.M. Femoral head size and wear of highly cross-linked polyethylene at 5 to 8 years. *Clin. Orthop.* **2009**, *467*, 3290–3296. [CrossRef] [PubMed]
58. Nakahara, I.; Nakamura, N.; Takao, M.; Sakai, T.; Nishii, T.; Sugano, N. Eight-year wear analysis in Longevity highly cross-linked polyethylene liners comparing 26- and 32-mm heads. *Arch. Orthop. Trauma Surg.* **2011**, *131*, 1731–1737. [CrossRef] [PubMed]
59. Cuckler, J.M.; Bearcroft, J.; Asgian, C.M. Femoral head technologies to reduce polyethylene wear in total hip arthroplasty. *Clin. Orthop.* **1995**, *317*, 57–63. [PubMed]
60. D'Antonio, J.A.; Sutton, K. Ceramic materials as bearing surfaces for total hip arthroplasty. *J. Am. Acad. Orthop. Surg.* **2009**, *17*, 63–68. [PubMed]
61. Galvin, A.L.; Jennings, L.M.; Tipper, J.L.; Ingham, E.; Fisher, J. Wear and creep of highly crosslinked polyethylene against cobalt chrome and ceramic femoral heads. *Proc. Inst. Mech. Eng.* **2010**, *224*, 1175–1183. [CrossRef]
62. Urban, J.A.; Garvin, K.L.; Boese, C.K.; Bryson, L.; Pedersen, D.R.; Callaghan, J.J.; Miller, R.K. Ceramic-on-polyethylene bearing surfaces in total hip arthroplasty. Seventeen to twenty-one year results. *J. Bone Joint Surg. Am.* **2001**, *83*, 1688–1694. [PubMed]
63. Kawate, K.; Omura, T.; Kawahara, I.; Tamai, K.; Ueha, T.; Takemura, K. Differences in highly cross-linked polyethylene wear between zirconia and cobalt-chromium femoral heads in Japanese patients: A prospective, randomized study. *J. Arthroplast.* **2009**, *24*, 1221–1224. [CrossRef] [PubMed]
64. Stilling, M.; Nielsen, K.A.; Soballe, K.; Rahbek, O. Clinical comparison of polyethylene wear with zirconia or cobalt-chromium femoral heads. *Clin. Orthop.* **2009**, *467*, 2644–2650. [CrossRef] [PubMed]
65. Kraay, M.J.; Thomas, R.D.; Rimnac, C.M.; Fitzgerald, S.J.; Goldberg, V.M. Zirconia *vs.* Co-Cr femoral heads in total hip arthroplasty: Early assessment of wear. *Clin. Orthop.* **2006**, *453*, 86–90. [CrossRef] [PubMed]

66. Meftah, M.; Klingenstein, G.G.; Yun, R.J.; Ranawat, A.S.; Ranawat, C.S. Long-term performance of ceramic and metal femoral heads on conventional polyethylene in young and active patients: A matched-pair analysis. *J. Bone Joint Surg. Am.* **2013**, *95*, 1193–1197. [CrossRef] [PubMed]
67. Laurent, M.P.; Johnson, T.S.; Crowninshield, R.D.; Blanchard, C.R.; Bambri, S.K.; Yao, J.Q. Characterization of a highly cross-linked ultrahigh molecular-weight polyethylene in clinical use in total hip arthroplasty. *J. Arthroplast.* **2008**, *23*, 751–761. [CrossRef] [PubMed]
68. ISO 14242-3-Implants for Surgery—Wear of Total Hip-Joint Prostheses—Part 3: Loading and displacement parameters for orbital bearing type wear testing machines and corresponding environmental conditions for test. International Organization for Standardization: Geneva, Switzerland, 2009.
69. F1714–96 (2013)-Standard Guide for Gravimetric Wear Assessment of Prosthetic Hip-Designs in Simulator Devices. ASTM International: West Conshohocken, PA, USA, 2013.
70. ISO 14242-2-Implants for Surgery—Wear of Total Hip-Joint Prostheses—Part 2: Methods of measurement. International Organization for Standardization: Geneva, Switzerland, 2000.
71. Jasty, M.; Rubash, H.E.; Muratoglu, O.K. Highly crosslinked polyethylene: The debate is over—In the affirmative. *J. Arthroplast.* **2005**, *20*, 55–58. [CrossRef]
72. Bragdon, C.R.; Barrett, S.; Martell, J.M.; Greene, M.E.; Henrik, M.; Harris, W.H. Steady state penetration rates of electron beam-irradiated, highly cross-linked polyethylene at an average of 45-month follow-up. *J. Arthroplast.* **2006**, *21*, 935–943. [CrossRef] [PubMed]

Article

Design of an Advanced Bearing System for Total Knee Arthroplasty

Mark L. Morrison [1,*], Shilesh Jani [2] and Amit Parikh [1]

[1] Smith and Nephew Advanced Surgical Devices, 1450 East Brooks Road, Memphis, TN 38116, USA; amit.parikh@smith-nephew.com

[2] Formerly Smith and Nephew, Currently Orchid Orthopedic Solutions, 4600 East Shelby Drive, Suite 1, Memphis, TN 38118, USA; shilesh.jani@orchid-ortho.com

[*] Author to whom correspondence should be addressed; mark.morrison@smith-nephew.com; Tel.: +1-901-399-5160; Fax: +1-901-399-6020.

Academic Editor: J. Philippe Kretzer

Received: 26 March 2015; Accepted: 29 May 2015; Published: 9 June 2015

Abstract: The objective of this study was to develop an advanced-bearing couple for TKA that optimizes the balance between wear resistance and mechanical properties. The mechanical and structural properties of virgin and highly crosslinked, re-melted UHMWPE were evaluated, and tibial inserts manufactured from these UHMWPE materials were tested against either oxidized zirconium (OxZr) or CoCr femoral components on a knee simulator. This study confirmed that the wear resistance of crosslinked UHMWPE improves with increasing radiation dose but is accompanied by a concomitant reduction in mechanical properties. Compared to CoCr, the ceramic surface of OxZr allows the use of a lower irradiation dose to achieve equivalent reductions in wear rates. As a result, a given wear rate can be achieved without sacrificing the mechanical properties to the same extent that is necessary with a CoCr femoral component. The advantage of ceramic counter bearing surfaces extends to both pristine and microabrasive conditions.

Keywords: highly crosslinked UHMWPE; knee; TKA; wear; mechanical properties; oxidized zirconium; abrasion; crosslink density

---

1. Introduction

The most common bearing couple used in total hip and knee arthroplasties (THA and TKA, respectively) is ultra-high molecular weight polyethylene (UHMWPE) articulating against a CoCr alloy. This couple has demonstrated excellent clinical results with survivorship greater than 90% at 13 years in THA [1] and greater than 92% at 13 years in TKA [1]. However, UHMWPE wear and subsequent osteolysis have been a primary long-term failure mechanism, which results in decreasing survivorship with increasing time *in vivo* [2–4]. Traditionally, THA and TKA were performed in elderly patients with the intent of reducing pain and providing mobility. However, a number of changes in patient demographics have occurred in the intervening decades, namely: (1) the population has become heavier; (2) patients are receiving THA and TKA at younger ages; and (3) the patients are more likely to continue active lifestyles after surgery [5–7]. Therefore, contemporary joint replacements experience greater biomechanical demands than experienced in the past, which create a need for more wear-resistant implants.

This need for improvement was first identified in THA due to the larger number of procedures performed, higher wear rates, and greater incidence of osteolysis. In the late 1990s, crosslinking, which was first used and quizzically abandoned in the 1970s [8], was resurrected as a technique for improved wear resistance of UHMWPE. Today, the formulations of crosslinked UHMWPE that were introduced into the THA marketplace in the late 1990s and early 2000s are showing promising clinical results with up to 12 years of follow-up [9–11]. Radiographic wear shows marked improvements, and the incidence of osteolysis secondary to wear shows dramatic reductions compared to conventional UHMWPE.

Historically, highly crosslinked UHMWPE did not see wide-spread use in TKA because crosslinking results in the reduction of particular mechanical properties and contact stresses can be higher in TKA than in THA. However, total knee replacement is now a more prevalent procedure and is predicted to grow at a faster rate than total hip replacement [7]. In addition, Sharkey *et al.* reported that one of the primary, long-term causes of TKA revision was polyethylene wear, which accounted for 25% of the failures in their retrospective review of 212 consecutive TKA revisions [12]. Ten years later, a follow-up study by Sharkey *et al.* [13] reported a remarkable decrease in revisions due to polyethylene wear and concluded that this reduction was, in part, likely due to the successful development of materials with improved wear resistance.

Clearly there is a need for further improvements in TKA bearings technology. Encouraged by the clinical success of crosslinked UHMWPE in the hip, there is now greater confidence that the material may also have a compelling application in knees. Engineers and clinicians, however, recognize that the stresses and motions in knees are different from those in hips and that the effects of the reduction in mechanical properties concomitant with crosslinking must be evaluated carefully. As a result, adoption of crosslinked UHMWPE in TKA has proceeded with greater caution.

It is well understood that the degree of crosslinking appropriate for a particular application is a delicate balance between improvements in wear and reduced mechanical properties resulting from increased crosslinking [14]. With that in mind, the crosslinked UHMWPE formulations currently marketed for TKA typically utilize lower crosslink densities than those marketed for THA.

The hypothesis of the present study is that the wear rates of various formulations of UHMWPE can be reduced not only by increasing crosslink density, but additionally by changing the counter bearing surface from the metallic CoCr alloy to a ceramic. Ceramics (typically sintered aluminum or zirconium oxides) are commonly utilized in THA femoral heads for two reasons, namely (1) reduced wear of UHMWPE because of a lower coefficient of friction compared to CoCr; and (2) improved microabrasion resistance because they are harder than CoCr. Both of these properties are singularly attributed to the predominantly covalently bonded molecular structure of ceramics. However, ceramics are brittle (due to the strong covalent bonds) and have found very limited use in the higher stress application of TKA. A solution to this limitation of bulk ceramics is to design a material system in which the component is fully ceramic only at the bearing surface and metallic in its core. Surface oxidized zirconium (OxZr), trademarked as OXINIUM™ (Smith and Nephew, Memphis, TN, USA), is one such material marketed for orthopaedic implants.

The objective of this study was to develop an advanced-bearing couple for TKA that optimizes the balance between wear resistance and mechanical properties. Wear testing was conducted under pristine and tumbled/abraded conditions to examine the performance and durability of the bearing couples under ideal and less-than-ideal conditions. Mechanical and structural characterizations of various UHMWPEs were also conducted as a function of crosslink density to rationalize the balance between wear resistance and mechanical/structural properties.

2. Materials and Methods

2.1. Materials

The UHMWPE materials examined in this study were fabricated from multiple lots of compression-molded GUR1020 UHMWPE rod stock that were obtained from MediTECH Medical Polymers (Fort Wayne, IN, USA). Some of the material was evaluated in the unirradiated state and is denoted as virgin material. Additional rods were crosslinked with nominal gamma-radiation doses of 5.0 Mrad (50 kGy), 7.5 Mrad (75 kGy), or 10.0 Mrad (100 kGy) and subsequently re-melted to stabilize them against oxidation. These highly crosslinked materials are identified as 5-XLPE, 7.5-XLPE and 10-XLPE, respectively. Only the components for wear testing were sterilized by ethylene oxide

(EtO) gas. The remainder of the material was evaluated in the unsterilized condition since multiple studies have demonstrated that EtO sterilization does not alter the properties of either virgin or highly crosslinked UHMWPE [15–17].

2.2. Mechanical and Physical Characterization

All of the mechanical and physical testing was conducted by MediTECH. Tensile testing (n = 5) was performed according to ASTM F648-07 and ASTM D638-03 with Type IV specimens with thicknesses of 3.0 ± 0.05 mm and displacement rates of 50 mm/min. Izod impact testing (n = 5) was conducted according to ASTM F648-07 and ASTM D256-05 with a 7.5 J hammer, and density (n = 3) was determined according to ASTM D792-00, Method B.

2.3. Thermophysical Analysis

Thermal analysis was conducted according to ASTM F2625-07 using a Netzsch 204 F1 Phoenix differential scanning calorimeter (DSC). Each DSC sample was removed from the core of the rod stock. Samples were cut and weighed to a resolution of 0.01 mg and ranged in mass from approximately 7 to 10 mg. An attempt was made to produce plate-like samples and keep the approximate dimensions consistent in order to minimize dimensional effects on the thermogram variability. The DSC cycle consisted of a 10 min equilibration at 30 °C, followed by heating to 180 °C at 10 °C/min rate, and cooling to 30 °C at 10 °C/min. For all of the thermograms, an empty aluminum crucible was used as the reference. First, an empty aluminum crucible was run as a baseline correction. The samples were crimped into aluminum crucibles and placed in the DSC chamber, which was continuously flushed with research-grade nitrogen gas at a flow rate of approximately 30 mL/min. Five samples (n = 5) were run per material condition with this heating profile. The resultant thermograms were analyzed to determine the extrapolated onset (T_{OM}) and peak melting (T_{PM}) temperatures and the heats of fusion (ΔH_m). For the determination of ΔH_m, the integration limits were systematically placed at 50 °C and 160 °C. The percent crystallinity (%X) was estimated as:

$$\%X = \frac{\Delta H_m}{291} \times 100 \quad (1)$$

where 291 J/g is the enthalpy associated with the melting of 100% crystalline polyethylene.

2.4. Molecular Network Parameters

The molecular network parameters were evaluated by Cambridge Polymer Group (Boston, MA, USA). Three cubes (n = 3) with nominal dimensions of 5 × 5 × 5 mm were prepared from each lot of material by Cambridge Polymer. It should be noted that the orientations of the samples relative to the compression axes can affect the results [18] but were unknown in this study. The virgin samples were evaluated according to ASTM D2765-01. The highly crosslinked samples were evaluated according to ASTM F2214-02, and the swell ratio (q_s) and crosslink density (ν_d) were reported. Two lots of virgin UHMWPE, 7 lots of 5-XLPE, 6 lots of 7.5-XLPE, and 3 lots of 10-XLPE were evaluated. The repeat measurements were averaged for each lot of material, and the replicate measurements for each lot were averaged for each level of crosslinking. Briefly, the initial heights (H_o) of the cube specimens were first measured using a digital micrometer with a resolution of 1 μm. The specimens were then immersed in o-xylene with 0.1 wt % Irganox™ antioxidant, and the specimen height was monitored with a lightweight, ceramic dilatometer probe until steady-state conditions were achieved (H_f = final height). As specified in the standard, the swell ratio was calculated according to Equation (2):

$$q_s = \left(\frac{H_f}{H_o}\right)^3 \quad (2)$$

Once q_s was determined, the crosslink density and molecular weight between crosslinks were calculated for all materials according to the following equations delineated in ASTM F2214:

$$\text{Crosslink Density} = \nu_d = \frac{Ln\left(1 - q_s^{-1}\right) + q_s^{-1} + \chi_1 q_s^{-2}}{\phi_1\left(q_s^{\frac{-1}{3}} - \frac{q_s^{-1}}{2}\right)} \quad (3)$$

$$\text{Molecular Weight Between Crosslinks} = M_c = \left(\overline{\nu}\nu_d\right)^{-1} \quad (4)$$

Where:

χ_1 = Heat of mixing for the polymer-solvent system (Flory interaction parameter) = $0.33 + 0.55q_s^{-1}$
Φ_1 = Molar volume of the solvent = 136 cm^3/mol
$\overline{\nu}$ = Specific volume of the polymer = 920 g/dm^3

2.5. Wear Resistance

The wear resistances of the virgin, 5-XLPE and 7.5-XLPE materials were evaluated on a knee simulator against pristine (*i.e.*, new) CoCr (ASTM F75) and OxZr [19] (ASTM F2384) femoral components. Per standard processing, the cast CoCr femorals were hot isostatic pressed (HIPed) followed by solution heat treating. In addition, testing was conducted with CoCr and OxZr femoral components that were subjected to a tumbling protocol to simulate the effects of microabrasion by third-body debris *in vivo* [20]. This protocol involved tumbling the femoral components in a centrifugal barrel mass-finisher for approximately 30 s in a 25 μm alumina powder and plastic cone media. These tumbled femoral components were then tested on a knee simulator against the various UHMWPE materials to evaluate the effect of the microabrasion on UHMWPE wear.

Genesis II™ (Smith and Nephew, Memphis, TN, USA) cruciate-retaining (CR) knee components (n = 3) were tested on a 6-station, displacement-controlled, knee simulator (AMTI, Watertown, MA, USA) and rotated weekly to minimize the effect of station variability. Unloaded soak controls (n = 3) were utilized to account for fluid uptake by the UHMWPE tibial inserts. The lubricant was alpha calf fraction (Hyclone Labs, Logan, UT) with 20 mM EDTA and 0.2% sodium azide, which was diluted to 50% with deionized water to achieve an average protein concentration of approximately 20 g/L. These tests were run for approximately 5 million cycles (Mc) each. The lubricant was replaced and the tibial inserts were weighed about every 0.5 Mc. The load/motion profiles were the "high kinematics" inputs described by Barnett *et al.* [21], which was based on a gait-lab study of young, healthy males by Lafortune [22]. In this study, the total A/P translation was 11 mm, which is slightly greater than that used by Barnett *et al.* The slope of the least squares best-fit line of cumulative volume loss *vs.* cycles was determined for each liner, averaged for each bearing combination and defined as the wear rate.

2.6. Roughness Measurements

Roughness measurements of the femoral components were made using a contact profilometer (Surfcom 1800D, Carl Zeiss, Brighton, MI, USA) with a 2 μm radius stylus tip, a 0.25 mm cut-off length and a 4.0 mm evaluation length. Roughness measurements (n = 10) were made systematically in the medial-lateral axis from 0° to 45° of flexion on each condyle of each femoral component before wear-simulator testing for both the pristine and tumbled components. Roughness was characterized using five parameters: R_a, the average surface displacement (of peaks and valleys) from the mean surface line; R_{pm}, the average peak height above the mean surface line; R_p, the maximum peak height above the mean surface line; R_{pk}, the average peak height above the mean surface line; and R_{sk}, the skewness, or asymmetry of the profile about the mean surface line.

2.7. Electron Spin Resonance

Evaluation of the residual free-radical concentrations in these materials was conducted to verify the effectiveness of the post-irradiation re-melting process in reducing the free radicals. Samples (n = 5) for electron spin resonance (ESR) with diameters of 3 mm and lengths of 10 mm were punched from the center of the rod stock for each material. A 7.5 Mrad gamma-irradiated material without any subsequent heat treatment was utilized as a positive control in this experiment.

The free-radical concentration (FRC) in each material was determined with an X-band ESR spectrometer (Bruker EMX300, University of Memphis) operating at a microwave frequency of approximately 9.8 GHz and employing a high-sensitivity universal X-band resonator cavity. ESR test parameters were constant for all measurements at 1 mW microwave power, 5 Gauss amplitude modulation, 200 G sweep width, 3500 center field, and a relatively high gain and Q-value. The absolute magnitude of the free-radical concentration was analyzed and computed with WinEPR analysis software. The area under the absorption curve of a test specimen was compared to that of the National Institute of Standards and Technology (NIST) intensity standard SRM-2601, and these results were normalized to the specimen's weight.

2.8. Delamination Evaluation

A unidirectional, reciprocating pin-on-disk wear test was conducted to evaluate the susceptibility of the 7.5-XLPE material to delaminate under worst-case conditions. Virgin, unirradiated UHMWPE was used as a negative control in this test because delamination in virgin polyethylene has not been observed clinically. Samples that were gamma sterilized in air were used as a positive control due to the well-known history of *in vivo* delamination for that material [16]. Three replicate samples (n = 3) were evaluated per material treatment. Prior to delamination testing, these samples were subjected to accelerated aging according to ASTM F2003-02 for 21 days.

Delamination testing was conducted on a multidirectional pin-on-disk device (OrthoPOD, AMTI, Watertown, MA, USA) with CoCr pins articulating against UHMWPE disks. The CoCr pins were fabricated with a tip radius of 19.81 mm (0.78 in) and polished to a fine, mirror-like finish. A load of 444.8 N (100 lb_f) was applied to each pin to produce an initial, maximum Herztian contact stress of 60.4 MPa, which is at the upper limit of contact stresses reported for TKA *in vivo* [23]. The stroke length was approximately 19 mm. These tests were conducted with reciprocating motion at a frequency of 1.24 Hz for a total of 3 million cycles in 100% alpha calf fraction serum (HyClone, Logan, UT, USA) that was changed at approximately every 500,000 cycles. At those times, the UHMWPE samples were rinsed and examined for signs of delamination with the naked eye. In the case of one gamma-air sample, delamination caused complete wear-through of the UHMWPE and resulted in the removal of this specific sample at less than 3 Mc.

Upon completion of 3 Mc, the samples were removed and any signs of delamination were noted. The samples were then sectioned along the long axis of the wear track, cut with a microtome to produce a relatively smooth cross-section, sputter-coated with gold, and examined with a scanning electron microscope (SEM) in secondary electron mode. Both the wear surface and the sub-surface were examined for any signs of cracking or delamination.

2.9. Statistical Analyses

Statistical analyses of the results were performed with Minitab 16 (Minitab, Inc., State College, PA, USA). The homogeneity of variances within the various data sets was evaluated with Levene's test. If the variances were homogeneous, either the Student's t-test with pooled variances or analysis of variance (ANOVA) with Tukey's *post hoc* test was conducted to determine statistically significant differences in the properties. If the variances were not homogeneous, either the Student's t-test with separate variances or Welch analysis of variance (ANOVA) with a Tukey's *post hoc* test was conducted to determine statistically significant differences in the properties. The level of significance (α) was 0.05.

3. Results and Discussion

3.1. Mechanical and Physical Characterization

While the yield strength (YS) exhibited small, but significant decreases ($p < 0.001$) of approximately 8%–9% upon irradiation to any dose (Table 1), the ultimate tensile strength (UTS) and elongation at break decreased linearly with increasing radiation dose with statistically significant differences between each dose (Figure 1). Compared to virgin, the UTS decreased 10%–21% ($p < 0.001$) and elongation decreased 27%–45% ($p < 0.001$) depending upon the selected irradiation dose (Figure 1b,c). Likewise, the impact strength decreased approximately 45%–59% upon crosslinking with statistically significant differences ($p < 0.001$) between each dose (Figure 1d).

Table 1. Summary of the mean (±standard deviations) properties evaluated in this study for compression-molded GUR1020 UHMWPE in the virgin and highly crosslinked, re-melted conditions.

Material Property	Nominal Radiation Dose (Mrad)			
	0.0	**5.0**	**7.5**	**10.0**
Yield Strength, YS (MPa)	22.6 ± 0.4	20.7 ± 0.3	20.8 ± 0.5	20.6 ± 0.2
Ultimate Tensile Strength, UTS (MPa)	54.9 ± 1.4	49.4 ± 2.3	46.9 ± 1.8	43.6 ± 2.5
Elongation at Break, EL (%)	516 ± 29	379 ± 16	323 ± 8	282 ± 15
Izod Impact Strength (kJ/m^2)	164 ± 13	90 ± 4	76 ± 2	67 ± 3
Density (g/cm^3)	0.936 ± 0.001	0.932 ± 0.001	0.932 ± 0.001	0.932 ± 0.002
Onset Melting Temperature, T_{OM} (°C)	125.9 ± 0.5	123.6 ± 0.4	122.2 ± 0.6	120.9 ± 1.0
Peak Melting Temperature, T_{PM} (°C)	140.8 ± 0.6	139.8 ± 0.4	141.2 ± 0.7	141.6 ± 0.4
Crystallinity (%)	55.9 ± 0.8	50.6 ± 0.6	50.7 ± 1.2	50.7 ± 0.9
Swell Ratio, q	17.57 ± 0.26	3.57 ± 0.10	3.08 ± 0.10	2.82 ± 0.17
Extract (%)	18.5 ± 3.5	-	-	-
Crosslink Density, ν_d (mol/dm^3)	0.011 ± 0.003	0.151 ± 0.008	0.204 ± 0.024	0.237 ± 0.018
Molecular Weight Between Crosslinks, M_c (10^3 g/mol)	90.55 ± 16.39	6.11 ± 0.31	4.62 ± 0.54	3.94 ± 0.30
Free-Radical Concentration, FRC (spins/g)	Not Detectable	Not Detectable	Not Detectable	Not Detectable
Maximum Oxidation Index, OI_{max}	0.02 ± 0.00	0.02 ± 0.01	0.01 ± 0.00	-

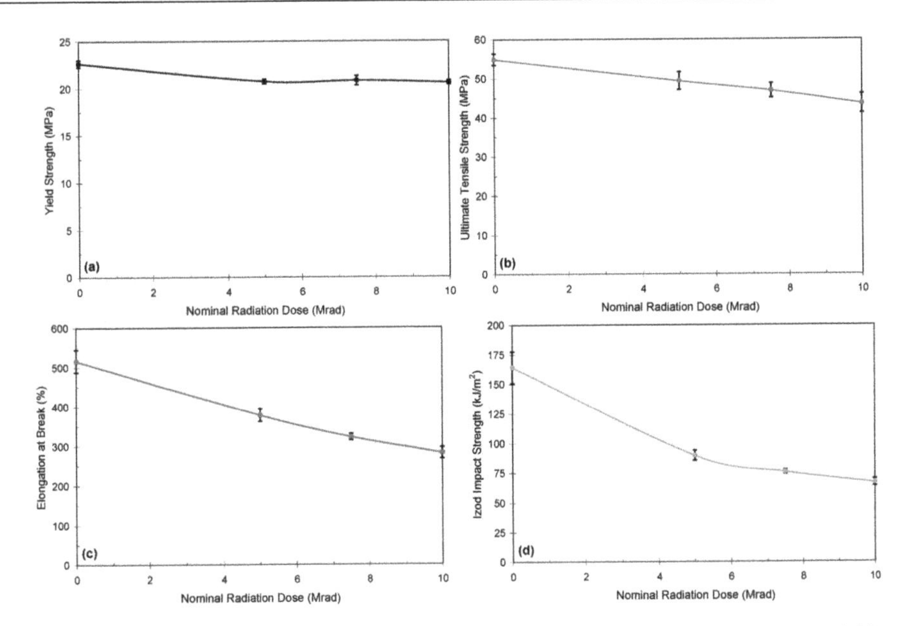

Figure 1. Plots of the **(a)** yield strength; **(b)** ultimate tensile strength; **(c)** elongation at break; and **(d)** Izod impact strength for GUR1020 UHMWPE irradiated with various gamma-radiation doses and subsequently re-melted.

3.2. Thermophysical Analysis

As demonstrated in Table 1, the T_{OM} was found to decrease by up to 4% in the crosslinked materials compared to the virgin materials with statistically significant differences ($p < 0.001$) for virgin compared to 7.5-XLPE and 10-XLPE and for 5-XLPE compared to 10-XLPE. The T_{PM} changed by $\pm 1\%$ or less in the crosslinked materials, relative to the virgin material, and some of these differences were statistically significant ($p < 0.001$). Finally, the %X decreased by approximately 10%, compared to virgin, upon irradiation to any of the irradiation doses examined in this study ($p < 0.001$).

3.3. Molecular Network Parameters

The mean swell ratio, percent extract, crosslink density, and molecular weight between crosslinks for the virgin and highly crosslinked materials are summarized in Table 1. The swell ratio (Figure 2) and molecular weight between crosslinks decreased with increasing radiation dose, and the crosslink density (Figure 3) increased with increasing radiation dose. The changes in all of these metrics were statistically significant ($p < 0.001$) as the dose increased to 7.5 Mrad. The differences in these metrics between the 7.5-XLPE and 10-XLPE materials was not significant at the levels evaluated in this study. Within the dose range examined, these values were similar to those reported by Greer and co-authors [24] as demonstrated in Figures 2 and 3.

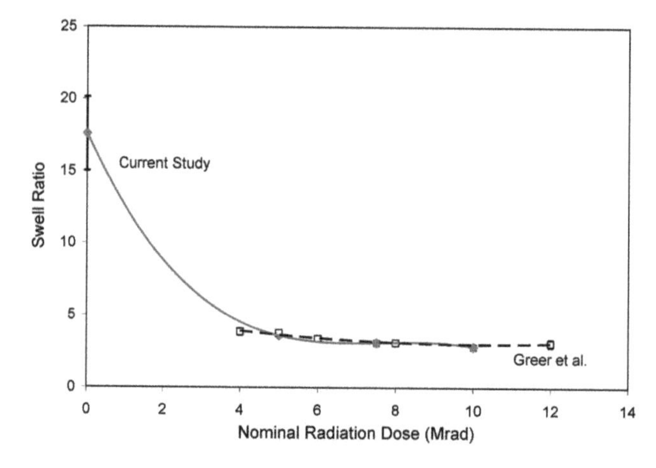

Figure 2. Plot of swell ratio for compression-molded GUR1020 UHMWPE in the virgin state and gamma-irradiated to various doses and re-melted. The results reported by Greer *et al.* [24] are also plotted for comparison. The error bars represent the standard deviations.

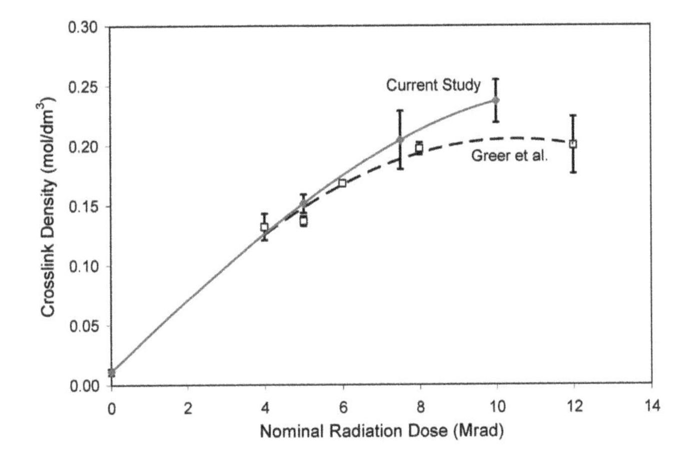

Figure 3. Plot of crosslink density for compression-molded GUR1020 UHMWPE in the virgin state and gamma-irradiated to various doses and re-melted. The results reported by Greer *et al.* [24] are also plotted for comparison. The error bars represent the standard deviations.

3.4. Roughness Measurements

The mean roughness parameters in the pristine conditions met the applicable quality specifications and standards for both femoral materials (Tables 2 and 3). After the CoCr femoral components were subjected to the tumbling protocol, all of the roughness parameters evaluated in this study increased ($p \leq 0.016$, Table 2). For the OxZr femorals (Table 3), all of the roughness parameters except for R_a (p = 0.062) and R_{sk} increased by statistically significant amounts ($p \leq 0.031$) after tumbling. In contrast, R_{sk} decreased significantly ($p < 0.001$) after tumbling. Comparing the materials in the tumbled conditions, all of the roughness parameters were significantly higher for CoCr relative to OxZr ($p < 0.001$).

Table 2. Summary of the mean (± standard deviations) roughness parameters for the CoCr femoral components in the pristine and roughened conditions. The levels of significance (p) for the differences between the pristine and roughened components are also reported.

Roughness Parameter	Pristine CoCr	Roughened CoCr	p
R_a (μm)	0.035 ± 0.004	0.077 ± 0.009	<0.001
R_{pm} (μm)	0.140 ± 0.013	0.510 ± 0.057	<0.001
R_p (μm)	0.216 ± 0.028	1.658 ± 0.160	<0.001
R_{pk} (μm)	0.066 ± 0.008	0.253 ± 0.022	<0.001
R_{sk}	0.618 ± 0.115	2.008 ± 1.363	0.016

Table 3. Summary of the mean (± standard deviations) roughness parameters for the OxZr femoral components in the pristine and roughened conditions. The levels of significance (p) for the differences between the pristine and roughened components are also reported.

Roughness Parameter	Pristine OxZr	Roughened OxZr	p
R_a (μm)	0.044 ± 0.007	0.050 ± 0.002	0.062
R_{pm} (μm)	0.152 ± 0.027	0.179 ± 0.011	0.001
R_p (μm)	0.248 ± 0.088	0.397 ± 0.066	0.019
R_{pk} (μm)	0.052 ± 0.009	0.072 ± 0.006	<0.001
R_{sk}	−1.443 ± 0.683	−2.801 ± 0.557	<0.001

3.5. *Wear Resistance*

Wear testing of pristine CoCr femoral components against virgin and highly crosslinked UHMWPE tibial inserts reflected documented trends in the literature [14] in that the mean wear rate decreased significantly ($p < 0.001$) with each increase in the radiation dose examined in this study (Figure 4). The mean wear rates (± standard deviations) of the virgin, 5-XLPE, and 7.5-XLPE tibial inserts were 23.4 ± 2.4 mm^3/Mc, 10.9 ± 0.9 mm^3/Mc, and 6.4 ± 0.6 mm^3/Mc, respectively. As a result, 5-XLPE produced a 53% reduction in wear while the use of 7.5-XLPE resulted in a 73% reduction in wear compared to virgin UHMWPE. Based on this trend with pristine CoCr femoral components, it is estimated that a nominal radiation dose of about 10 Mrad is necessary to produce a wear rate that is not measurable by the gravimetric technique.

Knee simulator wear testing of pristine OxZr femoral components against virgin, 5-XLPE, and 7.5-XLPE tibial inserts yielded wear rates of 11.7 ± 1.9 mm^3/Mc, 3.1 ± 1.0 mm^3/Mc, and 1.4 ± 0.2 mm^3/Mc, respectively. As a result, 5-XLPE produced a 73% reduction in wear while the use of 7.5-XLPE resulted in an 88% reduction in wear compared to virgin UHMWPE against pristine OxZr femoral components. Statistical analysis demonstrated that the wear rates for the highly crosslinked materials were significantly lower than that of the virgin material ($p < 0.001$), but there was insufficient evidence of differences between the 5-XLPE and 7.5-XLPE at the levels tested in this study. This data suggests that an estimated, nominal radiation dose of about 8 Mrad is necessary to produce a wear rate that is not measurable by the gravimetric technique when paired with pristine OxZr (Figure 4). More importantly, pristine OxZr femoral components alone reduced wear by 50% to 79% ($p \leq 0.001$) compared to pristine CoCr femoral components tested against the same UHMWPE materials.

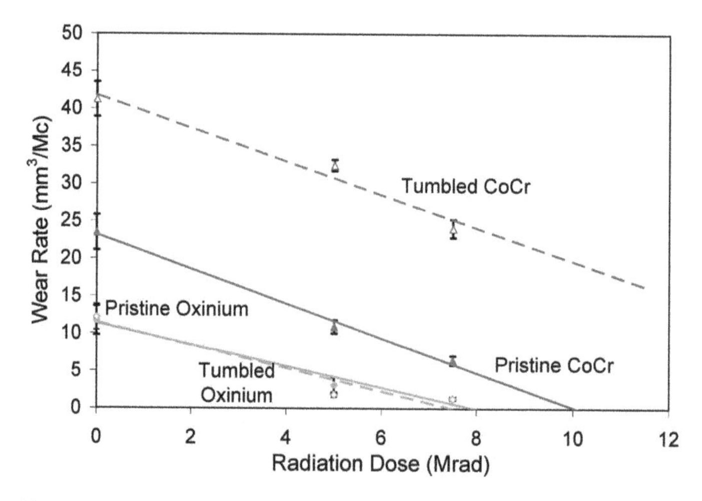

Figure 4. Plot of the mean wear rates (±standard deviations) in a knee simulator for UHMWPE crosslinked to various doses against either CoCr or OxZr femoral components in pristine (solid symbols and lines) and tumbled (open symbols and dashed lines) conditions. The regression lines have been extrapolated to radiation doses greater than 7.5 Mrad.

When the CoCr femoral components were tumbled to simulate *in vivo* microabrasion of the surfaces due to third-body debris, the wear rates for each material increased by statistically significant amounts ($p \leq 0.001$) compared to those calculated against pristine CoCr femoral components (Figure 4). In these tests, the mean wear rates (± standard deviations) of virgin, 5-XLPE, and 7.5-XLPE tibial inserts were 41.2 ± 2.3 mm^3/Mc, 32.5 ± 0.7 mm^3/Mc, and 24.0 ± 1.2 mm^3/Mc, respectively. This data demonstrates that, although the trend between radiation dose and wear rate is similar, the dramatic improvements in wear resistance for highly crosslinked materials are degraded by abrasion of the CoCr femoral components.

In contrast, no statistically significant changes in wear rates ($p \geq 0.113$) were observed, relative to pristine OxZr, when each material was tested against OxZr femoral components that were exposed to the same microabrasive challenge (Figure 4). For these simulator tests, the mean wear rates (±standard deviations) of virgin, 5-XLPE, and 7.5-XLPE tibial inserts were 12.2 ± 1.7 mm^3/Mc, 1.8 ± 0.4 mm^3/Mc, and 1.3 ± 0.1 mm^3/Mc, respectively. Under these microabrasive conditions, the wear rates against tumbled OxZr remained 49% to 83% lower ($p \leq 0.003$) than the wear rates observed for pristine CoCr femoral components tested against the same UHMWPE materials.

3.6. Electron Spin Resonance

The virgin, 5-XLPE, 7.5-XLPE, and 10-XLPE specimens did not show detectable ESR signals, indicating that the free-radical concentration (Table 1) was below the sensitivity of the instrument, which is approximately 1×10^{13} spins/g. In contrast, the 7.5 Mrad irradiated, unannealed material (positive control) exhibited a high mean FRC (±SD) of $2.36 \times 10^{16} \pm 0.36 \times 10^{16}$ spins/g, which is 3 orders of magnitude greater than the instrument sensitivity.

3.7. Delamination Study

The virgin and highly crosslinked samples endured 3 Mc of testing without any evidence of cracking or delamination on the wear surfaces (Figure 5a,b) or on the longitudinal cross-sections of the wear tracks in the post-test examinations. In contrast, all of the gamma-air-sterilized samples (Figure 5c) exhibited the onset of delamination prior to the completion of the full 3 Mc test protocol.

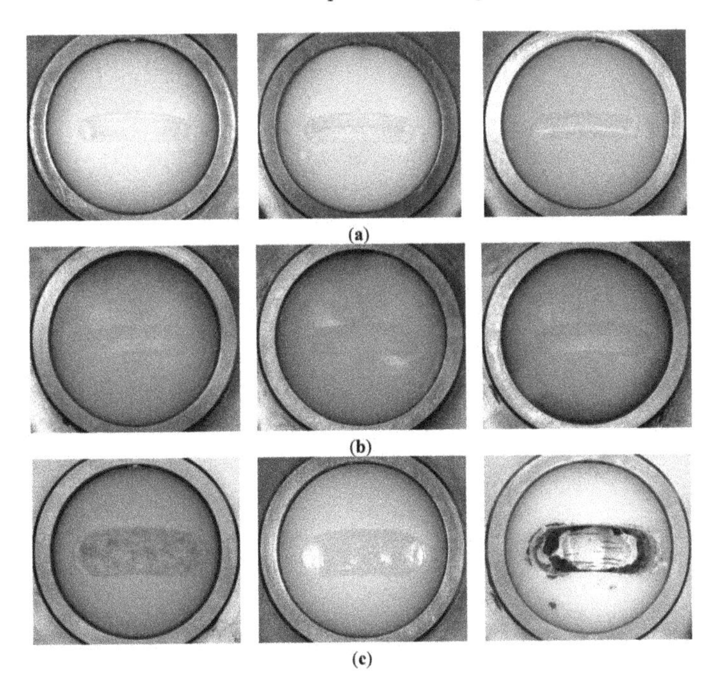

Figure 5. Photomacrographs of the (**a**) virgin; (**b**) 7.5-XLPE; and (**c**) gamma-air-sterilized samples after 3 million cycles of reciprocating pin-on-disk testing. Although wear tracks are visible on the virgin and 7.5-XLPE samples, no cracking or delamination was observed.

In this pin-on-disk study, the highly crosslinked UHMWPE materials did not exhibit a propensity for delamination under worst-case testing conditions. In addition, this particular test protocol was validated by the absence of delamination in the negative control and mild to extensive delamination in the positive controls, which replicated the damage observed in clinical retrievals. The resistance of highly crosslinked and re-melted UHMWPE to delamination has been similarly demonstrated by others in both pin-on-disk [25] and knee simulator testing [26].

4. Discussion

This study confirmed that increasing crosslink density (*i.e.*, increasing the radiation dose) results in significant reductions in wear rates. This trend was independent of the femoral bearing material. When the femoral component was pristine CoCr alloy, the wear rate *versus* radiation dose trend suggests that a crosslink dose of 10 Mrad would result in wear rates that are not measurable by the gravimetric technique. This does not mean that the wear rate is zero, rather that the wear rate is low enough that gravimetric measurement errors cannot distinguish it from zero as has been previously reported for hip simulator testing on 10 Mrad irradiated and re-melted UHMWPE [27].

This study also confirmed that the ceramic bearing surface of oxidized zirconium resulted in greater reductions in wear of crosslinked UHMWPE than a pristine CoCr bearing surface. When the relationships between wear rate and radiation dose for pristine CoCr and ceramic oxidized zirconium are compared, it was found that oxidized zirconium was more effective in reducing the wear of UHMWPE at a given radiation dose. This finding was expected considering the widely documented advantage of ceramic bearing materials when tested against UHMWPE in both THA [28,29] and TKA [30–32]. Subsequent knee-simulator testing has demonstrated that the low wear provided by OxZr femorals paired with 7.5-XLPE tibial inserts is maintained for an unprecedented 45 Mc of simulator testing [33,34].

The factorial (or percent) differences in UHMWPE wear rates between CoCr and OxZr increase as the crosslinking dose increases; 50% for virgin UHMWPE and 79% for 7.5 Mrad UHMWPE. However, the absolute difference in wear rates between CoCr and OxZr diminishes as the crosslink dose increases; 11.7 mm^3/Mc for virgin *vs.* 5 mm^3/Mc for 7.5 Mrad UHMWPE. The ramification of these trends is that a lower crosslinking dose can be used with OxZr than with CoCr to achieve a given wear reduction. For instance, in order to achieve a wear rate of approximately 3 mm^3/Mc, one would need to use a crosslinking dose of about 6 Mrad with OxZr *vs.* approximately 9 Mrad with CoCr. This additional 3 Mrad of radiation has consequences to the mechanical and structural properties of UHMWPE, as will be discussed later.

The discussion thus far has focused on knee simulator wear testing of femoral components in the pristine condition. When the femoral components were subjected to a microabrasive tumbling protocol, as expected the metallic, CoCr surfaces roughen to a greater extent than the harder [35], ceramic surfaces of OxZr. After tumbling, the roughness metrics measured on the CoCr components were well within the range reported previously for clinical retrievals [36]. Because the ceramic surface of OxZr is twice as hard as CoCr (12.1 GPa and 5.5 GPa, respectively, based on nanoindentation per the Oliver-Pharr method [35]), it was expected to roughen to a lesser degree than CoCr. However, the finding that the wear rates of UHMWPE against OxZr were unaffected by the tumbling protocol was unexpected, because OxZr did, indeed, exhibit increases in the roughness parameters. Understanding this observation requires a more detailed review of the roughness data.

The tumbling of CoCr and OxZr resulted in increases in the average surface roughness (R_a) for both materials. However, the increase in R_a was greater for CoCr (+122%, $p < 0.001$) than for OxZr (+12%, $p = 0.062$) relative to the respective pristine materials. Similar trends were observed for R_p, R_{pm}, and R_{pk}. The surface skewness (R_{sk}) exhibited a different behavior and illustrates the primary difference between metallic and ceramic counterfaces. CoCr femorals almost always exhibited positive values of R_{sk}, which means that there were more peaks than valleys on the bearing surface. After the CoCr femorals were subjected to the tumbling protocol, the scratches resulted in a greater quantity of larger peaks due to pile-up on the sides of the scratches, and the R_{sk} generally increased to greater positive values (+225%, $p = 0.016$). In contrast, the OxZr femorals always exhibited negative values of R_{sk}, which means that there are more valleys than peaks. After the tumbling protocol, R_{sk} decreased (−94%, $p < 0.001$) to more negative values due to scratching of the surface (*i.e.*, valleys) with little pile-up (*i.e.*, peaks), which is characteristic of ceramic bearing surfaces [37].

The wear rates of each of the UHMWPE formulations increased when CoCr was roughened compared to the corresponding wear rates under pristine conditions. During articulation of these components, the hard peaks on the CoCr femorals plow into the softer UHMWPE tibial inserts and generate greater amounts of wear. When UHMWPE was tested against OxZr subjected to the same microabrasive challenge, the resulting knee simulator wear rates were not elevated compared to the corresponding wear rates under pristine conditions. This lack of increase in wear after tumbling is primarily due to the smoother counterface with negative skewness. In other words, the tumbled OxZr femorals do not have many peaks that can plow into the UHMWPE tibial inserts and generate wear debris. As a result, wear does not increase even after a microabrasive challenge.

When these tumbled components were tested in a knee simulator against virgin and the various formulations of highly crosslinked UHMWPEs, the trends in wear rates were similar to those noted above for pristine femoral components. Increasing the crosslink dose resulted in improved wear resistance against both femoral materials. However, the wear rates increased when CoCr was abraded, while they did not change when OxZr was tumbled. If one wanted to achieve a wear rate of approximately 3 mm^3/Mc with tumbled OxZr as the counter bearing surface, a crosslink dose of about 6 Mrad would be required. To achieve that same 3 mm^3/Mc wear rate with abraded CoCr as the counter bearing surface, a predicted crosslink dose of approximately 18 Mrad would be required. Such a high crosslink dose is unlikely to be useful clinically because of the attendant loss of mechanical properties.

The wear-test results discussed above have highlighted the following trends: (1) the wear rates decrease with increasing radiation dose (crosslink density); (2) this trend is similar for both of the femoral bearing materials evaluated; (3) this trend is independent of the surface roughness of the counter bearing surface; (4) ceramic bearing surfaces result in lower UHMWPE wear rates than CoCr alloy bearing surfaces independent of the crosslinking dose; (5) wear rates of all formulations of UHMWPE increase markedly when CoCr counter bearing surfaces are abraded; and (6) wear rates of all formulations of UHMWPE are unaffected when OxZr ceramic counter bearing surfaces are subjected to a microabrasive challenge.

This study also confirmed previous studies [24,38,39] that showed that mechanical properties are affected by radiation crosslinking, particularly ultimate tensile strength (UTS), elongation at break, and impact strength. It stands to reason that the mechanically strongest material would be the ideal choice for use in orthopaedic implants. However, the mechanically strongest material (virgin, non-crosslinked UHMWPE) is also the least resistant to wear. Therefore, a balance between wear resistance and mechanical properties needs to be achieved and should also take into account the design of the orthopaedic implant. As discussed above, the ability to use a ceramic counter bearing surface affords greater reduction in wear rates with a given level of crosslinking as compared to CoCr surfaces. This study demonstrated that ceramic surfaces provide reductions in wear that are equivalent to an additional 3–5 Mrad crosslinking dose compared to pristine CoCr. When CoCr is roughened, this advantage of ceramics is elevated by an estimated 12 Mrad of crosslinking dose.

The crosslinked UHMWPE materials tested in this study were all re-melted after irradiation, and did not undergo any further irradiation. Because of this thermal treatment, all crosslinked formulations were found to have no measurable free radicals, which improves the resistance to oxidation [40]. However, it should be noted that UHMWPE that is prone to oxidation is also prone to *in vivo* degradation, particularly subsurface delamination wear [16]. This study confirmed that virgin (non-irradiated) and crosslinked, re-melted UHMWPEs are resistant to delamination wear after accelerated aging. Under identical conditions, the UHMWPE material that was irradiated in air, and subsequently aged showed copious delamination as has been documented clinically.

In summary, this study confirmed that crosslinking of UHMWPE for improved wear resistance comes at the expense of reduced mechanical properties. For applications in TKA, a majority of the research to-date has focused on striking the right balance between wear resistance and mechanical properties through selection of the irradiation dose alone. This study, however, showed that the counter bearing surface can have a profound effect on the wear rates of UHMWPEs. Compared to CoCr, the ceramic surface of OxZr allows the use of a lower irradiation dose to achieve equivalent reductions in wear rates. As a result, a given wear rate can be achieved without sacrificing the mechanical properties to the same extent that is necessary if a CoCr femoral component is used. The advantage of ceramic counter bearing surfaces extends to both pristine and microabrasive conditions.

5. Conclusions

This study confirmed that the wear resistance of crosslinked UHMWPE in TKA improves with increasing irradiation dose. However, along with this advantage is the concomitant reduction in key mechanical properties in highly crosslinked UHMWPE. When CoCr is used as the counter bearing surface, the appropriate balance between the wear resistance and the mechanical properties of crosslinked UHMWPE can be controlled only by the choice of the irradiation dose within a narrow range. On the other hand, when a ceramic counter bearing surface (e.g., oxidized zirconium) is used in the bearing couple under ideal, pristine conditions, the resultant improvement in wear resistance is equivalent to an additional 3–5 Mrad of radiation without sacrificing additional UHMWPE mechanical properties. This improvement in wear resistance is expanded even further to the equivalent of an approximate dose of 12 Mrad under microabrasive conditions.

The oxidized zirconium ceramic bearing surface articulating against UHMWPE crosslinked to a dose of 7.5 Mrad can provide a wear resistance that is similar to pristine CoCr articulating against approximately 10 Mrad crosslinked UHMWPE. If microabrasion of the CoCr femoral occurs *in vivo*, this difference in wear resistance grows even greater. Furthermore, the mechanical properties of the UHMWPE are preserved due to the lower irradiation dose that is necessary to achieve a given wear rate against oxidized zirconium.

Finally, re-melting of the radiation crosslinked UHMWPE results in elimination of all measurable free radicals and imparts oxidative stability to the material and attendant resistance to delamination wear.

Acknowledgments: Gary Hines and Carolyn Weaver are gratefully acknowledged for their assistance in this study.

Author Contributions: Mark Morrison, Shilesh Jani and Amit Parikh conceived and designed the experiments; Mark Morrison and Amit Parikh performed the experiments and analyzed the data; Mark Morrison, Shilesh Jani and Amit Parikh wrote the paper.

Conflicts of Interest: All of the authors were employees of Smith and Nephew at the time of this study.

References

1. Australian Orthopaedic Association National Joint Replacement Registry. *Annual Report*; Australian Orthopaedic Association: Adelaide, Australia, 2014.
2. Archibeck, M.J.; Jacobs, J.J.; Roebuck, K.A.; Glant, T.T. The basic science of periprosthetic osteolysis. *J. Bone Jt. Surg. Am.* **2001**, *82*, 1478–1489.
3. Harris, W.H. Wear and periprosthetic osteolysis: The problem. *Clin. Orthop. Relat. Res.* **2001**, *393*, 66–70. [CrossRef] [PubMed]
4. Gupta, S.K.; Chu, A.; Ranawat, A.S.; Slamin, J.; Ranawat, C.S. Review article: Osteolysis after total knee arthroplasty. *J. Arthroplast.* **2007**, *22*, 787–799. [CrossRef] [PubMed]
5. D'Apuzzo, M.R.; Hernandez-Polo, V.H.; Sierra, R.J. *National Trends in Primary Total Knee Arthroplasty: A Population-Based Study*; American Academy of Orthopaedic Surgeons: New Orleans, LA, USA, 2010; p. 681.

6. Dahl, A.W.; Robertsson, O.; Lidgren, L. *Surgical Treatment for Knee OA in Younger Patients*; American Academy of Orthopaedic Surgeons: New Orleans, LA, USA, 2010; p. 126.
7. Crowninshield, R.D.; Rosenberg, A.G.; Sporer, S.M. Changing demographics of patients with total joint replacement. *Clin. Orthop. Relat. Res.* **2006**, *443*, 266–272. [CrossRef] [PubMed]
8. Kurtz, S.M.; Ong, K. Contemporary total hip arthroplasty: Hard-on-hard bearings and highly crosslinked UHMWPE. In *UHMWPE Biomaterials Handbook*, 2nd ed.; Kurtz, S.M., Ed.; Academic Press: Burlington, MA, USA, 2009; pp. 55–79.
9. Epinette, J.A.; Manley, M.T. No differences found in bearing related hip survivorship at 10–12 years follow-up between patients with ceramic on highly cross-linked polyethylene bearings compared to patients with ceramic on ceramic bearings. *J. Arthroplast.* **2014**, *29*, 1369–1372. [CrossRef] [PubMed]
10. Glyn-Jones, S.; Thomas, G.E.R.; Garfjeld-Roberts, P.; Gundle, R.; Taylor, A.; McLardy-Smith, P.; Murray, D.W. The John Charnley award: Highly crosslinked polyethylene in total hip arthroplasty decreases long-term wear: A double-blind randomized trial. *Clin. Orthop. Relat. Res.* **2015**, *473*, 432–438. [CrossRef] [PubMed]
11. Engh, C.A., Jr.; Hopper, R.H., Jr.; Huynh, C.; Ho, H.; Sritulanondha, S.; Engh, C.A., Sr. A prospective, randomized study of cross-linked and non–cross-linked polyethylene for total hip arthroplasty at 10-year follow-up. *J. Arthroplast.* **2012**, *27*, 2–7. [CrossRef] [PubMed]
12. Sharkey, P.F.; Hozack, W.J.; Rothman, R.H.; Shastri, S.; Jacoby, S.M. Insall award paper. Why are total knee arthroplasties failing today? *Clin. Orthop. Relat. Res.* **2002**, *404*, 7–13. [CrossRef] [PubMed]
13. Sharkey, P.F.; Lichstein, P.M.; Shen, C.; Tokarski, A.T.; Parvizi, J. Why are total knee arthroplasties failing today—Has anything changed after 10 years? *J. Arthroplast.* **2014**, *29*, 1774–1778. [CrossRef] [PubMed]
14. Asano, T.; Akagi, M.; Clarke, I.C.; Masuda, S.; Ishii, T.; Nakamura, T. Dose effects of cross-linking polyethylene for total knee arthroplasty on wear performance and mechanical properties. *J. Biomed. Mater. Res. B* **2007**, *83*, 615–622. [CrossRef] [PubMed]
15. Muratoglu, O.K.; Merrill, E.W.; Bragdon, C.R.; O'Connor, D.; Hoeffel, D.; Burroughs, B.; Jasty, M.; Harris, W.H. Effect of radiation, heat, and aging on *in vitro* wear resistance of polyethylene. *Clin. Orthop. Relat. Res.* **2003**, *417*, 253–262. [PubMed]
16. Williams, I.R.; Mayor, M.B.; Collier, J.P. The impact of sterilization method on wear in knee arthroplasty. *Clin. Orthop. Relat. Res.* **1998**, *356*, 170–180. [CrossRef] [PubMed]
17. Goldman, M.; Lee, M.; Pruitt, L.; Gronsky, R. *A Comparison of Sterilization Techniques on the Structure and Morphology of Medical Grade UHMWPE*; Society for Biomaterials: Toronto, ON, Canada, 1996; p. 189.
18. Terrill, L.; Liao, C.W.; Lin, S.; DeAzevedo, M.; Sun, D.C. *Effects of Anisotropy in UHMWPE on Crosslinked Density Testing*; Orthopaedic Research Society: San Francisco, CA, USA, 2008; p. 1674.
19. Sheth, N.P.; Lementowski, P.; Hunter, G.; Garino, J.P. Clinical applications of oxidized zirconium. *J. Surg. Orthop. Adv.* **2008**, *17*, 17–26. [PubMed]
20. Widding, K.; Hines, G.; Hunter, G.; Salehi, A. *Knee Simulator Protocol for Testing of Oxidized Zirconium and Cobalt Chrome Femoral Components under Abrasive Conditions*; Orthopaedic Research Society: Dallas, TX, USA, 2002; p. 1009.
21. Barnett, P.I.; Fisher, J.; Auger, D.D.; Stone, M.H.; Ingham, E. Comparison of wear in a total knee replacement under different kinematic conditions. *J. Mater. Sci. Mater. Med.* **2001**, *12*, 1039–1042. [CrossRef] [PubMed]
22. Lafortune, M.A.; Cavanagh, P.R.; Sommer, H.J., III; Kalenak, A. Three-dimensional kinematics of the human knee during walking. *J. Biomech.* **1992**, *25*, 347–357. [CrossRef]
23. Bartel, D.L.; Rawlinson, J.J.; Burstein, A.H.; Ranawat, C.S.; Flynn, W.F., Jr. Stresses in polyethylene components of contemporary total knee replacements. *Clin. Orthop. Relat. Res.* **1995**, *317*, 76–82. [PubMed]
24. Greer, K.W.; King, R.S.; Chan, F.W. The effects of raw material, irradiation dose, and irradiation source on crosslinking of UHMWPE. In *Crosslinked and Thermally Treated Ultra-High Molecular Weight Polyethylene for Joint Replacements, ASTM STP 1445*; Kurtz, S.M., Gsell, R., Martell, J., Eds.; ASTM International: West Conshohocken, PA, USA, 2003; pp. 209–220.
25. Burroughs, B.; Muratoglu, O.K.; O'Connor, D.O.; Bragdon, C.R.; Harris, W.H. *Development of a Pitting/Delamination Pin on Disc Wear Model to Simulate the Failure Modes Often Seen in UHMWPE Tibial Knee Inserts*; Orthopaedic Research Society: Orlando, FL, USA, 2000; p. 0552.
26. Deluzio, K.J.; Muratoglu, O.K.; O'Connor, D.O.; Bragdon, C.R.; O'Flynn, H.; Rubash, H.; Jasty, M.; Wyss, U.P.; Harris, W.H. *Development of an in vitro Knee Delamination Model in a Knee Simulator with Physiologic Load and Motion*; World Biomater Cong: Kamuela, HI, USA, 2000; p. 842.

27. Good, V.; Ries, M.; Barrack, R.L.; Widding, K.; Hunter, G.; Heuer, D. Reduced wear with oxidized zirconium femoral heads. *J. Bone Jt. Surg. Am.* **2003**, *85A*, 105–110.
28. Dahl, J.; Nivbrant, B.; Soderlund, P.; Nordsletten, L.; Rohrl, S.M. Less wear with aluminumoxide heads against conventional polyethylene—A 10 year follow-up radiostereometry (RSA). *J. Bone Jt. Surg. Br.* **2010**, *92*, 392–397.
29. Wroblewski, B.M.; Siney, P.D.; Dowson, D.; Collins, S.N. Prospective clinical and joint simulator studies of a new total hip arthroplasty using alumina ceramic heads and cross-linked polyethylene cups. *J. Bone Jt. Surg. Br.* **1996**, *78*, 280–285.
30. Kaddick, C.; Streicher, R. Ceramic total knee replacements: Do they produce less wear? *Bone Jt. J.* **2013**, *95*, 94–97.
31. Oonishi, H.; Ueno, M.; Kim, S.C.; Oonishi, H.; Iwamoto, M.; Kyomoto, M. Ceramic *vs.* cobalt-chrome femoral components; wear of polyethylene insert in total knee prosthesis. *J. Arthroplast.* **2009**, *24*, 374–382. [CrossRef] [PubMed]
32. Zietz, C.; Bergschmidt, P.; Lange, R.; Mittelmeier, W.; Bader, R. Third-body abrasive wear of tibial polyethylene inserts combined with metallic and ceramic femoral components in a knee simulator study. *Int. J. Artif. Org.* **2013**, *36*, 47–55. [CrossRef] [PubMed]
33. Papannagari, R.; Hines, G.; Sprague, J.; Morrison, M. *Long-Term Wear Performance of an Advanced Bearing Knee Technology*; ISTA: Dubai, UAE, 2010; pp. A12–A13.
34. Papannagari, R.; Hines, G.; Morrison, M.; Sprague, J. *Long-Term Wear Performance of an Advanced Bearing Technology for TKA*; Orthopaedic Research Society: Long Beach, CA, USA, 2011; p. 1141.
35. Long, M.; Riester, L.; Hunter, G. *Nano-Hardness Measurements of Oxidized Zr-2.5Nb and Various Orthopaedic Materials*; Society Biomater: San Diego, CA, USA, 1998; p. 528.
36. Levesque, M.; Livingston, B.J.; Jones, W.M.; Spector, M. *Scratches on Condyles in Normal Functioning Total Knee Arthroplasty*; Orthopaedic Research Society: New Orleans, LA, USA, 1998; pp. 241–247.
37. Cooper, J.R.; Dowson, D.; Fisher, J.; Jobbins, B. Ceramic bearing surfaces in total artificial joints: Resistance to third body wear damage from bone cement particles. *J. Med. Eng. Technol.* **1991**, *15*, 63–67. [CrossRef] [PubMed]
38. Pruitt, L.A. Deformation, yielding, fracture and fatigue behavior of conventional and highly cross-linked ultra high molecular weight polyethylene. *Biomaterials* **2005**, *26*, 905–915. [CrossRef] [PubMed]
39. Oral, E.; Malhi, A.S.; Muratoglu, O.K. Mechanisms of decrease in fatigue crack propagation resistance in irradiated and melted UHMWPE. *Biomaterials* **2006**, *27*, 917–925. [CrossRef] [PubMed]
40. Muratoglu, O.K. Highly crosslinked and melted UHMWPE. In *UHMWPE Biomaterials Handbook*, 2nd ed.; Kurtz, S.M., Ed.; Academic Press: Burlington, MA, USA, 2009; pp. 197–204.

Review

In Vitro Analyses of the Toxicity, Immunological, and Gene Expression Effects of Cobalt-Chromium Alloy Wear Debris and Co Ions Derived from Metal-on-Metal Hip Implants

Olga M. Posada [1,4,*], Rothwelle J. Tate [2], R.M. Dominic Meek [3] and M. Helen Grant [1]

1. Biomedical Engineering Department, University of Strathclyde, Wolfson Centre, Glasgow G4 0NW, UK; m.h.grant@strath.ac.uk
2. Strathclyde Institute for Pharmacy & Biomedical Sciences, University of Strathclyde, Glasgow G4 0RE, UK; r.j.tate@strath.ac.uk
3. Department of Orthopaedic Surgery, Southern General Hospital, Glasgow G11 6NT, UK; rmdmeek@doctors.org.uk
4. Current affiliation, Leeds Institute of Cardiovascular and Metabolic Medicine (LICAMM), University of Leeds, Leeds LS2 9JT, UK

* Author to whom correspondence should be addressed; medomp@leeds.ac.uk; Tel.: +44-113-343-7747.

Academic Editor: J. Philippe Kretzer
Received: 30 April 2015; Accepted: 8 July 2015; Published: 14 July 2015

Abstract: Joint replacement has proven to be an extremely successful and cost-effective means of relieving arthritic pain and improving quality of life for recipients. Wear debris-induced osteolysis is, however, a major limitation and causes orthopaedic implant aseptic loosening, and various cell types including macrophages, monocytes, osteoblasts, and osteoclasts, are involved. During the last few years, there has been increasing concern about metal-on-metal (MoM) hip replacements regarding adverse reactions to metal debris associated with the MoM articulation. Even though MoM-bearing technology was initially aimed to extend the durability of hip replacements and to reduce the requirement for revision, they have been reported to release at least three times more cobalt and chromium ions than metal-on-polyethylene (MoP) hip replacements. As a result, the toxicity of metal particles and ions produced by bearing surfaces, both locally in the periprosthetic space and systemically, became a concern. Several investigations have been carried out to understand the mechanisms responsible for the adverse response to metal wear debris. This review aims at summarising *in vitro* analyses of the toxicity, immunological, and gene expression effects of cobalt ions and wear debris derived from MoM hip implants.

Keywords: wear debris; cobalt; metal-on-metal; hip implants

1. Introduction

Joint replacement has proven to be an extremely successful and cost-effective means of relieving arthritic pain and improving quality of life for recipients [1]. A primary replacement is an initial replacement procedure undertaken on a joint and involves replacing either part (partial) or all (total) of the articular surface. Revision hip replacements are repeat-operations of previous hip replacements where one or more of the prosthetic components are replaced, removed, or one or more components are added [2].

The total number of joint replacement procedures recorded by the National Joint Registry of England, Wales, and Northern Ireland (NJR) exceeded 1.6 million records between 1 April 2003 and 31 March 2013, with 2012/13 having the highest ever annual number of submissions at 205,686. The total number of primary hip procedures was 620,400. Of these, 76,274 were entered into the NJR during 2013, as reported in the NJR 11th Annual Report [3]. Similarly, there have been 410,767 hip replacements reported to the Australian Registry up to 31 December 2013. Of these, 40,180 were entered during 2013 [2].

Total joint replacement surgery has traditionally been reserved for elderly patients with advanced arthrosis who postoperatively would be less active. However, this scenario has now substantially changed, and many patients now receive total hip arthroplasty at a younger age. For example, Furnes *et al.* [4] compared implant survival of metal-on-metal (MoM) with that of metal-on-highly-cross-linked-polyethylene in patients between 45 and 64 years old. Kanda *et al.* [5] reported the case of a 42-year-old patient presenting with femoral head migration after an arthroplasty performed 22 years earlier. Moreover, it has been estimated that 10,000 to 30,000 patients less than 25 years of age have undergone joint replacement procedures in the last five years, and it is likely that many of those are paediatric patients [6]. It can also be anticipated that the rate of joint replacement in paediatric patients will increase, particularly given the popularity of this surgery and the incidence of diagnoses that may result in joint replacement surgery. Currently, over 294,000 individuals younger than 21 years of age are estimated to have juvenile arthritis [7]. The clinical outcome is generally excellent, but many young patients still need implant replacement within 10–15 years [8], and some may experience complications, including implant failure. Out of the 620,400 procedures recorded by the NJR, 14,903 had an associated first revision. The most commonly cited indications were aseptic loosening (cited in 3659 procedures), pain (3489), dislocation/subluxation (2545), and infection (2072) [3]. In accordance with this, the Australian registry reported the most common reasons for revision of primary total conventional hip replacement were loosening/lysis (2550 procedures), followed by prosthesis dislocation (2251), fracture (1576), and infection (1534) [2].

Wear debris-induced osteolysis is a major cause of orthopaedic implant aseptic loosening, and various cell types including macrophages, monocytes, osteoblasts, and osteoclasts are involved [9]. During the last few years, there has been increasing concern about MoM hip replacements regarding adverse reactions to metal debris associated with the MoM articulation [10]. As a consequence, on 22 July 2008, there was a voluntary recall of the Zimmer Durom® Acetabular Component ("Durom Cup") because the instructions for use/surgical technique were inadequate, which led to a higher than expected revision rate. Following this, on 24 August 2010, there was a voluntary recall of the DePuy ASR™ total hip system because of new, unpublished data from the UK joint registry indicating the revision rates within five years were approximately 13%. Two years later, on 1 June 2012, Smith & Nephew Orthopaedics initiated a market withdrawal for metal liners of the R3 acetabular system due to a higher than expected number of revision surgeries associated with the use of the device in total hip replacements outside the US [11]. Since a patient with an adverse reaction to metal debris can be asymptomatic [12], may have low metal ion levels [13], and may have normal cross-sectional imaging [14], diagnosing an adverse reaction is challenging. Even though MoM-bearing technology was initially aimed to extend the durability of hip replacements and to reduce the requirement for revision, they have been reported to release at least three times more cobalt and chromium ions than metal-on-polyethylene (MoP) hip replacements [15,16]. As a result, the toxicity of metal particles and ions produced by bearing surfaces, both locally in the periprosthetic space and systemically, became a concern [17]. Several investigations have been carried out to understand the mechanisms responsible for the adverse response to metal wear debris. This review aims at summarising *in vitro* analyses of the toxicity, immunological, and gene expression effects of cobalt ions and wear debris derived from MoM hip implants.

2. Wear Particle Generation and Metal Ion Release

The degradation products of any orthopaedic implant include two basic types of debris: particles and soluble (or ionic) debris [18]. MoM joints are exposed to a complex *in vivo* environment with mechanical and electrochemical degradation mechanisms that influence the longevity of the device [19].

The wear of MoM joints is of particular concern because particulate debris and release of Co/Cr ions can lead to adverse tissue reactions including necrosis, hypersensitivity, and pseudotumors [20,21].

Wear characteristics of MoM total hip replacements have two distinctive stages. Initially, the femoral and acetabular components show a relatively rapid but decreasing wear rate over the first 1–2 $\times$ 10^6 cycles in a hip joint simulator, or for one or two years *in vivo*, generally referred to as the "bedding-in" or "running-in" stage. Once this process has been completed, the rate of wear becomes reasonably steady and hence is referred to as "steady-state" [22,23]. Most wear debris particles are created in the running-in phase [23]. Wear debris is released from the articular surfaces after joint arthroplasty as a result of friction between articulating implant components or between cement and implant [24]. There are three types of mechanical wear mechanisms: fatigue, abrasive, and adhesive [9]. The first is caused by cyclic stresses inducing micro-fractures to occur within materials due to fatigue; once these micro-fractures reach the surface, wear particles are generated through delamination [25]. Abrasive wear can be split into two sub-categories; two-body and three-body. Two-body abrasive wear involves the roughness of a hard surface in contact with a softer material; particulates are released from the softer material due to ploughing. Three-body abrasive wear involves three materials instead; for example, they can include bone cement or fragments of bone between two articulating surfaces. Finally, adhesive wear involves intermolecular bonds of the weaker material bonding to the stronger material, resulting in shearing [9].

Metal corrosion is the degradation process affecting the surface of metallic materials due to their reaction with the surrounding environment. Most metallic materials are susceptible to corrosion attack if a tenacious surface oxide layer does not exist. Where the surface layer is permeable to oxygen and moisture, the corrosion process will continue and lead to eventual failure. Among the variety of corrosion mechanisms, metal corrosion is driven mainly by electrochemical potential. During exposure to aqueous environments, atoms of the metal surface experience an anodic process; electrons are released from the atoms forming metallic ions (oxidation). The localized electrical potential accelerates the oxidation process until the electrochemical potential is balanced [26]. Fretting corrosion is a type of mechanically assisted chemical degradation. Fretting corrosion damage is determined by a combination of metal atom dissolution through the fractured passive layer and metal oxide reformation. The oxide film is fractured by the contact and friction of rough articulating surfaces and exposure of the pure metal surface to a corrosive medium [27,28]. The physiological environment is considered corrosive. This makes the corrosion of metallic materials a slow and continuous process, which leads to the release of metal ions [29,30]. Corrosion damage is a very important issue for metallic implants that can affect their biocompatibility and mechanical integrity [31]. Chloride ions, amino acids, and proteins in the body can accelerate corrosion. Metallic biomaterials in aqueous solutions comprise active and passive surfaces that are simultaneously in contact with electrolytes. In this environment, the surface oxide film repeats a process of partial dissolution and re-precipitation. Metal ions are gradually released when dissolution is faster than re-precipitation [32]. Under physiological conditions, corrosion occurs as an electrochemical process in which electron exchange occurs at the metal surface [33]. The rate of this phenomenon is determined, in part, by the surface area. Since wear debris released from metal components is, for the most part, in the nanometre size range, it has a high surface area that increases the rate of corrosion [28]. When CoCr alloy is in contact with body fluids, cobalt is completely dissolved, and the surface oxide changes into chromium oxide containing a small amount of molybdenum oxide [34]. Elevated levels of Co and Cr ions occur in the peripheral blood and in the hip synovial fluid after CoCr alloy MoM hip replacement, and there is concern about the toxicity and biological effects of such ions both locally and systemically [35,36] (Table 1).

It has been reported that 20%–30% material loss can be attributed to corrosion-related damage, and that not only includes the pure corrosion process but also the wear induced/enhanced corrosion process that is defined as tribocorrosion [23]. Mechanically assisted corrosion, also referred to as tribocorrosion, is an irreversible process that occurs on the surface causing a deterioration of the material due to the combined wear and corrosion actions that simultaneously take place [31]. Tribocorrosion, present at bearing surfaces and within modular taper connections between components of the arthroplasty device, has been proposed to be the primary process by which ions and particles are generated [37]. The released ions can activate the immune system by forming complexes with native proteins [38]. Chromium and cobalt have similar protein binding affinity, and bind to proteins in proportion to the concentration ratio [39]. Once a metal is bound to a protein, it can be systemically transported and either stored or excreted [28]. Additionally, the presence of wear debris in the peri-implant area leads to macrophage phagocytosis of particulate debris and activation to stimulation of the release of a variety of mediators, such as free radicals and nitric oxide, and a myriad of proinflammatory cytokines and chemokines [40]. It has been reported that local acidification may develop during acute and chronic inflammation [41] and high hydrogen ion concentrations down to pH 5.4 have been found in inflamed tissue [42]. In turn, such an acidic environment, created by actively metabolizing immune cells, may enhance the corrosion process, increasing the metal ions being released. To illustrate this, Posada *et al.* [43] showed that significantly higher concentrations of Co and Cr were released when CoCr metal wear debris were incubated at low pH (Figure 1). Their findings suggest that the osteolysis process generated by wear debris may be exacerbated by the lowering of pH at an inflammation site. This is in line with reports of synovial fluid acidosis correlating with radiological joint destruction in rheumatoid-arthritis knee joints [44,45].

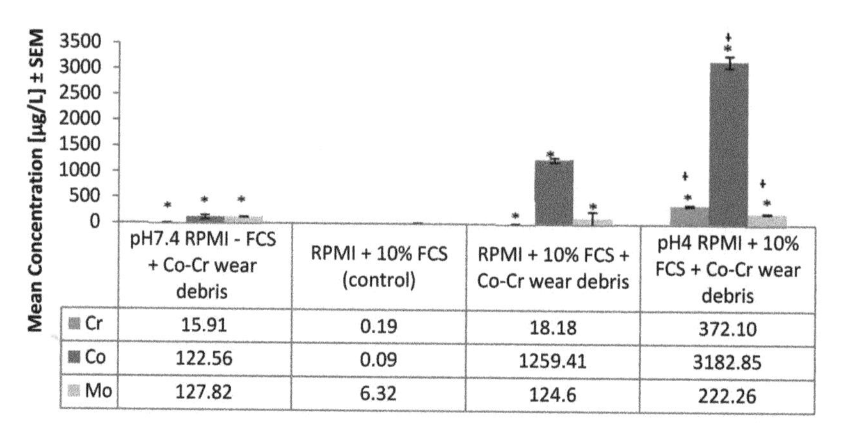

	pH7.4 RPMI - FCS + Co-Cr wear debris	RPMI + 10% FCS (control)	RPMI + 10% FCS + Co-Cr wear debris	pH4 RPMI + 10% FCS + Co-Cr wear debris
Cr	15.91	0.19	18.18	372.10
Co	122.56	0.09	1259.41	3182.85
Mo	127.82	6.32	124.6	222.26

Figure 1. Metal ions released *in vitro* from CoCr alloy into RPMI-1640 medium at pH 7.4 and 4.0. Results are expressed as mean values (±standard error of the mean (SEM), $n = 3$). FCS = foetal calf serum. * Significantly different from control values ($p < 0.05$) by one-way Analysis Of Variance (ANOVA) followed by Dunnett's multiple comparison test. Significant difference between pH 7.4 and pH 4.0.

2.1. Chromium

After entering the body from an exogenous source, Cr^{3+} binds to plasma proteins such as transferrin, an iron-transporting protein. Regardless of the source, Cr^{3+} is widely distributed in the body and accounts for most of the chromium in plasma or tissues. The Cr^{3+} is taken up as a protein complex into bone marrow, lungs, lymph nodes, spleen, kidney, and liver, with the highest uptake being in the lungs [46]. It has been shown that cell membranes are relatively impermeable to Cr^{3+}. When varying amounts of radioactive Cr^{3+} were added to whole blood *in vitro*, almost all of the radioactivity (94%–99%) remained in the plasma with an insignificant amount retained in the red blood cells (RBC). Similar results were obtained *in vivo* [47]. Similarly, low permeability of Cr^{3+} was found in Chinese hamster lung V79 cells exposed to Cr^{3+} complexes [48]. Additionally, it has been shown that the cellular uptake of Cr^{6+} is several-fold greater than that of Cr^{3+} ion, because trivalent chromium is predominantly octahedral and diffuses slowly [49]. In contrast to Cr^{3+}, Cr^{6+} is rapidly taken up by RBCs and reduced to Cr^{3+} inside the cell. Cr^{6+} enters the cell through non-specific anionic channels, such as the phosphate and sulphate anion exchange pathway [50,51]. Once within the cell, Cr^{6+} is reduced metabolically by the redox system to short-lived intermediates Cr^{5+}, Cr^{4+}, and ultimately to the most stable species Cr^{3+} [52–54]. Cr^{3+} interacts and forms complexes with DNA, protein, and lipids resulting in increased chromium intracellular levels [51–53].

Table 1. Metal ion levels measured in whole blood and synovial fluid from patients with metal-on-metal (MoM) hip replacements.

Author	Implant	Body fluid	Follow up	Mean Concentration (µg/L)		
					Co	Cr
Daniel, *et al.* [55]	MoM resurfacing	Whole blood	Up to 4 years	Pre-op	0.2	0.3
				1 year	1.3	2.4
				4 years	1.2	1.1
Ziaee, *et al.* [56]	MoM resurfacing	Whole blood	Mean of 53 months	Control	0.3	0.2
				Patients	1.4	1.9
Antoniou, *et al.* [57]	MoM (THA and resurfacing), MoP (THA)	Whole blood	1 year	Control	1.8	0.1
				MoM THA	2.6	0.6
				MoM resurfacing	2.4	0.5
				MoP THA	1.7	0.1
				non-steep (component abduction <55°)	2.4	3.6
Wretenberg [58]	MoM THA (Case report)	Whole blood	37 years	-	22.9	19.4
Hart, *et al.* [59]	Painful MoM resurfacings	Whole blood	median of 27 months	Unilateral	4.5 *	3.0 *
				Bilateral	10.6 *	7.9 *
Hart, *et al.* [60]	Failed MoM resurfacing	Whole blood	Mean 51 months	-	112.6	61.7
Langton, *et al.* [61]	MoM resurfacing (ASR, BHR)	Whole blood	minimum of 12 months	ASR	2.7 *	4.2 *
				BHR	1.8 *	4.2 *
				Adverse reactions	69.0	29.3
Davda, *et al.* [62]	Symptomatic MoM, THA and resurfacing	Synovial fluid	Mean of 36 months	Unexplained pain	1127.0 *	1337.0 *
				Defined cause of failure	1014.0 *	1512.0 *
Hart, *et al.* [63]	MoM, THA and resurfacing	Whole blood	39–42 months	Failed	6.9 *	5.0 *
				Well-functioning	1.7 *	2.3 *
				Non-pseudotumor	1.9 *	2.1 *
				Pseudotumor	9.2 *	12.0 *
Malviya, *et al.* [64]	MoM, MoP, THA	Whole blood	2 years	MoM	5.2	2.8
				MoP	1.6	0.8
Fritzsche, *et al.* [65]	bilateral MoM resurfacing followed by unilateral MoM THA (Case report)	Whole blood, aspirate of pseudotumor	3 months after revision surgery	Blood	138.0	39.0
				Aspirate of pseudotumor	258.0	1011.0
				Well-functioning	2.3	1.6
Matthies, *et al.* [66]	MoM, THA and resurfacing	Whole blood	Median of 39 months	No pseudotumor	2.9	3.2
				Pseudotumor	11.0	6.7
Lass, *et al.* [67]	MoM, THA	Synovial fluid	minimum of 18 years	-	113.4 *	54.0 *

* Ion concentrations expressed as median values. Pre-op = previous to operation, THA = total hip arthroplasty, MoM = metal-on-metal, MoP = metal-on-polyethylene, BHR = Birmingham Hip Resurfacing, ASR = Acetabular System Resurfacing, DePuy.

Excretion of Cr occurs primarily via urine, with no major retention in organs. Approximately 10% of an absorbed dose is eliminated by biliary excretion, with smaller amounts excreted in hair, nails, milk, and sweat. Clearance from plasma is generally rapid (within hours), whereas elimination from tissues is slower (with a half-life of several days) [68].

The toxicity, mutagenicity, and carcinogenicity of chromium compounds are well-established phenomena [46,50,69,70]. Long-term occupational inhalational exposure to Cr levels 100–1000 times higher than those found in the natural environment have been associated with squamous cell carcinoma and adenocarcinoma in exposed workers [71]. Epidemiological studies carried out in the UK, Europe, Japan, and the United States have consistently shown that workers in occupations where particulate chromates are generated or used have an elevated risk of respiratory disease, fibrosis, perforation of the nasal septum, development of nasal polyps, and lung cancer [72]. Additionally, during the intracellular reduction of Cr^{6+} to the stable Cr^{3+}, reactive intermediates (for example, reactive oxygen species (ROS), pentavalent and tetravalent chromium species) are generated, which causes a wide variety of DNA lesions including Cr-DNA adducts, DNA-protein crosslinks, DNA-DNA crosslinks, and oxidative damage [68,73].

Cr toxicity is associated with its oxidation state. However, it is still controversial whether Cr is released as Cr^{6+} in patients with MoM devices, with some reports supporting this idea and others disproving it [60,70,74]. In the authors' experience, analysis of the speciation of Cr is fraught with difficulty due to the instability of Cr^{6+}, which tends to oxidise to Cr^{3+} very rapidly. There is a general consensus, however, that Cr(III) is elevated in the biological fluids of all patients with MoM-type implants [63,75].

2.2. Cobalt

For the general population, diet is the main source of exposure to Co and it is readily absorbed from the small intestine [68,76]. Most of the consumed Co is excreted in the urine and the little that is retained is mainly in the liver and kidneys [68]. Under physiological conditions, this element is mostly accumulated in the liver, kidneys, heart, and spleen, while minimum concentrations are found in the blood serum and tissues of the brain and pancreas [77]. Molecular details of the mode of Co uptake into cells are not well known [76,78]. However, it is likely that it is transported into the cells by broad-specificity divalent metal transporters [78]. It has been shown that P2X7, a transmembrane ionotropic receptor, is involved in the uptake of divalent cations and Co [79]. In the same way, a protein named divalent metal transporter 1 (DMT1) has been shown to have a broad substrate specificity favouring divalent metals including Co^{2+} [80–82]. Additionally, the cellular uptake of Co may be mediated both by active transport ion pumps (*i.e.*, Ca^{2+}/Mg^{2+} ATPase and Na^{+}/K^{+} ATPase) and endocytosis [84].

The only biological known function of Co is as an integral part of vitamin B12, which is incorporated into enzymes that participate in reactions essential to DNA synthesis, fatty acid synthesis, and energy production [76,78]. Even though Co has a role in biological systems, overexposure results in toxicity [78], which involves development of hypoxia, increases in the level of ROS, suppression of Adenosine triphosphate (ATP) synthesis, and initiation of apoptotic and necrotic cell death [77]. Co ions can directly induce DNA damage, interfere with DNA repair, and cause DNA-protein crosslinking and sister chromatid exchange [68]. The exact mechanism for Co carcinogenicity remains to be elucidated, but it has been established that free radical generation contributes to its toxicity and carcinogenicity [83].

3. Biological Effects

3.1. Toxicity

Particulate wear debris generated by MoM has an average particle size range of 30 to 100 nm [84]. The reduced size of the particles allows their entry into tissues and organs and diffusion throughout the body, and interaction with different types of cells [85]. Concern about the toxicity of such particles has led to a number of studies assessing the effects of CoCr metal wear debris *in vitro* on a variety of cells [86–88]. It has been established that both Co ions and Co nanoparticles are cytotoxic and induce apoptosis and, at high concentrations, they induce necrosis with an inflammatory response [89].

Papageorgiou *et al.* [86] compared the cytotoxic and genotoxic effects of nanoparticles and micron-sized particles of CoCr alloy using human fibroblasts. Their results showed that exposure to both nano- and micron-sized particles of CoCr alloy, at the same particle mass per cell, causes different types and amounts of cellular damage. Posada *et al.* [88] investigated the effects of the combined exposure to CoCr nanoparticles and cobalt ions released from a resurfacing implant on monocytes (U937 cells), and used much lower concentrations of nanoparticles than the previous study [85]. They showed that metal debris in combination with Co ions had a direct effect on cell viability. Interestingly, they showed that previous exposure to Co ions seems to sensitise U937 cells to the toxic effects of both Co ions themselves and to nanoparticles, pointing to the potential for interaction *in vivo*. Their results indicate that even low doses of CoCr nanoparticles can exert cytotoxic effects. Dalal *et al.* [87] compared the responses of human osteoblasts, fibroblasts, and macrophages exposed to different metal-based particles (*i.e.*, cobalt-chromium (CoCr) alloy, titanium (Ti) alloy, zirconium (Zr) oxide, and Zr alloy). They found that CoCr-alloy particles were by far the most toxic and decreased viability and proliferation of human osteoblasts, fibroblasts, and macrophages. VanOs *et al.* [90] used commercially available 60 nm and 700 nm round chromium oxide (Cr^2O^3) particles to analyse the cytotoxic effects of chromium oxide particles on macrophage responses *in vitro*. With both particle sizes, cell mortality increased, resulting in a significant decrease in total cell numbers, as well as a significant increase in late apoptosis and necrosis. Tsaousi *et al.* [91] investigated the *in vitro* cytotoxicity and genotoxic effects of alumina ceramic (Al_2O_3) particles in comparison with CoCr alloy particles. They found no significant differences in cell viability between control and ceramic-treated cells, at all doses and time-points studied. However, and in agreement with the studies mentioned above, cells exposed to CoCr alloy particles showed both dose- and time-dependent cytotoxicity including damage to, and loss of, chromosomes.

The apoptotic effects of Co ions have mainly been reported at concentrations starting from 100 μM, where Co induced cell death and apoptosis in a dose- and time-dependent manner [92–94]. Catelas *et al.* [95] demonstrated that macrophage mortality induced by metal ions depends on the type and concentration of metal ions as well as the duration of their exposure. Overall, apoptosis was predominant after 24 h with both Co^{2+} (0–10 ppm) and Cr^{3+} (0–500 ppm) ions, but high concentrations induced mainly necrosis at 48 h. This same group also showed that Co^{2+} and Cr^{3+} induced mortality and apoptosis in J774 macrophages [96,97]. In a similar way, Akbar *et al.* [94] reported that exposure to high concentrations of metal ions (10 and 100 μM Cr^{6+}, 100 μM Co^{2+}) initiated apoptosis that resulted in decreased lymphocyte proliferation. A variety of soluble metals, including Co^{2+} and Cr^{3+}, at a range of concentrations between 0.05 and 5 mM, were found to induce Jurkat T-lymphocyte DNA damage, apoptosis, and/or direct necrosis in a metal- and concentration-dependent manner [98].

From all these reports, it seems evident that CoCr nanoparticles and metal ions released from MoM implants have toxic effects *in vitro* and may pose a health risk to patients, regardless of whether their implant is well-functioning or failing. This toxicity helps explain the higher prevalence of adverse reactions to metal debris when compared to ceramic or polyethylene particles. The long-term effects of the exposure to these particles and ions remain a concern. Adverse health effects caused by accumulated metal particles in the periprosthetic tissues include osteolysis [99], inflammation, pain, and pseudotumours [100]. Case *et al.* [101] reported that the accumulation of metal particles in lymph nodes causes structural changes such as necrosis and fibrosis. Multiple reports [102–109] have described patients with MoM implants who presented systemic adverse effects including neurological symptoms such as auditory impairment/deafness, visual impairment/blindness, peripheral neuropathy/dysesthesia of the extremities, poor concentration/cognitive decline, cardiomyopathy, and hypothyroidism. Several authors have associated these adverse effects with grossly elevated systemic Co blood levels. For example, Devlin *et al.* [110] reviewed 10 cases of suspected prosthetic hip-associated cobalt toxicity and reported that these patients had findings consistent with cobalt toxicity, including thyroid, cardiac, and neurologic dysfunction. Similarly, Bradberry *et al.* [111] reviewed some cases in which patients exposed to high circulating concentrations of cobalt from failed hip replacements developed neurological damage, hypothyroidism, and/or cardiomyopathy. Finally, Clark *et al.* [112] reported that chronic exposure to MoM hip resurfacing is associated with subtle structural changes in the visual pathways and the basal ganglia in asymptomatic patients. Consistent with this notion, revision surgery to remove the defective metal hip prostheses resulted in lowered blood concentrations of metal ions and improved symptoms. Evidence is accumulating that systemic elevated concentrations of Co ions, due to the presence of wear debris, pose a serious health risk for some patients bearing CoCr MoM implants. There has also been speculation of a potential carcinogenic effect, however, recent reports [113,114] suggest that CoCr-containing hip implants are unlikely to be associated with an increased risk of cancer.

3.2. *Immunological*

Wear debris products generated at the articulation site may lead to a chronic inflammatory reaction in the periprosthetic region, resulting in implant failure caused by macrophage-stimulated osteolysis and aseptic loosening [115,116], which is the principal biological mechanism underlying prosthesis failure according to the National Joint Registry of England, Wales, and Northern Ireland [117].

It has been established and accepted that the presence of implant devices and wear debris incites a foreign body inflammatory reaction [118]. Metallic debris derived from alloy implants induces macrophage activation and triggers immune responses resulting in the release of an array of proinflammatory mediators including Tumor necrosis factor alpha (TNF)-α, IL-1β, IL-6, and IL-8 [119]. TNF-α, IL-1, and IL-6 induce osteoclast differentiation and maturation, which lead to bone resorption and, ultimately, aseptic loosening [120] (Figure 2). Dalal *et al.* [87] reported an increase in TNF-α and IL-8 production by human osteoblasts, fibroblasts, and macrophages in response to different metal-based particles. Interestingly, they observed that the greatest cytokine responses of macrophages were to CoCr alloy particles. Posada *et al.* [88] also reported higher levels of secretion of IL-6, TNF-α, and interferon (IFN)-γ by resting monocyte-like cells (U937) after exposure to high concentrations of metal debris and the combination of metal debris and Co ions. Devitt *et al.* [121] investigated the *in vitro* effect of Co ions on a variety of cell lines by measuring production of IL-8 and Monocyte chemoattractant protein-1 (MCP-1) and found that Co ions enhanced the secretion of both cytokines in renal epithelial cells, gastric and colon epithelium, monocytes and neutrophils, and small airway epithelial cells. These investigations suggest a key role of Co ions in the immune response to wear debris.

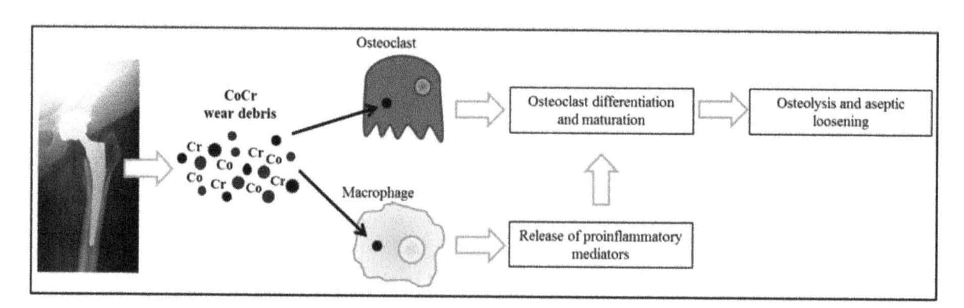

Figure 2. Schematic diagram of macrophage-stimulated osteolysis.

Despite the understanding of implant-related cytokine/chemokine networks that are released by different peri-implant cell types, knowledge about the mechanisms mediating cellular interaction with debris particles and the subsequent activation of macrophages to produce and release the inflammatory mediators remain incomplete [122]. This has led to a number of studies seeking to define such mechanisms. Toll-like receptors (TLRs) are mainly found on monocytes and macrophages, and have been previously shown to activate the inflammatory cascade by triggering the expression of various cytokines (IL-1, IL-6, TNF-a), growth factors (macrophage colony stimulating factor-1), and chemokines (MIP-1 a, MCP-1), and activating various downstream signalling pathways (nuclear factor-κB (NF-κB), protein kinase B (AKT/PKB), and mitogen-activated protein kinase (MAPK) [123]. Tyson-Capper *et al.* [124] demonstrated that Co ions released from MoM joint replacement implants stimulate innate immune responses via direct activation of TLR4. Similarly, Potnis *et al.* [125] also demonstrated that Co-alloy particles trigger immune responses via the Toll-like receptor 4 (TLR4) myeloid differentiation primary response protein 88 (MyD88)-dependent signalling pathway. Since a key element in the initiation of the innate immune response against pathogens is the recognition of components commonly found on the pathogen, referred to as pathogen-associated molecular patterns (PAMP) [126], these studies suggest that wear particles could have a PAMP-like behaviour and bind to TLRs being expressed by macrophages, which then initiates signalling pathways leading to the stimulation of the immune response. Additionally, endogenous molecules generated upon tissue injury, termed damage-associated molecular patterns (DAMPs), directly activate TLRs [127]. Proteins released by the damaged periprosthetic tissue include several heat shock proteins, biglycan, and fragments of extracellular molecules, such as oligosaccharides of hyaluronic acid and heparan sulphate, all of which are known activators of TLR2 and TLR4 [1].

Metal debris and ions can activate the immune system by inducing a delayed-type hypersensitivity reaction [38]. The most common sensitizing orthopaedic metals are nickel, cobalt, and chromium [38,128–130]. It is thought that stimulated T-cells generate pro-osteoclastogenetic factors that can alter bone homeostasis [115] and therefore contribute to osteolysis. The prevalence of metal sensitivity among the general population is approximately 10% to 15% and the prevalence of metal sensitivity among patients with well-functioning and poorly functioning implants has been reported to be ~25% and 60%, respectively, as measured by dermal patch testing [38]. The response of metal-specific lymphocytes to metals as debris or in the form of metal ions has been linked to poor implant performance. Cell-mediated type-IV hypersensitivity reaction characterized by vasculitis with perivascular and intramural lymphocytic infiltration of the postcapillary venules, swelling of the vascular endothelium, recurrent localized bleeding, and necrosis has been reported following MoM hip replacements [131]. Lymphocyte infiltrates have also been reported in the soft-tissue masses, described as pseudo-tumours, following MoM resurfacing arthroplasty [21,132]. Additionally, metal-specific T-cells have been isolated from patients with contact dermatitis, indicating a T-cell-led inflammatory reaction against a metal-derived antigen [133]. For example, Cr exposure has been shown to upset the immunoregulatory balance between Th1 and Th2 cells that controls different immune effects or functions through the production of distinct cytokines [134]. In the work of these authors, the effects of cadmium, chromium, inorganic mercury, and inorganic lead exposure on the immune system were determined by measuring cytokine production of human peripheral blood mononuclear cells. Their results showed that the cytokines assayed were differentially affected by heavy metal exposure. Of particular interest, Cr significantly increased the production of IL-1β while decreasing the production of IFN-γ, IL-6, IL-8, and IL-10.

It has been suggested that activation of T-cells following exposure to biomaterial particles is driven by macrophages and requires synergistic signals primed by both antigen presentation and costimulation [135–137]. Bainbridge *et al.* [138] examined the expression of CD80 and CD86 costimulatory molecules in U937 cells that had been exposed to titanium aluminium vanadium alloy (TiAlV) implant wear debris. This was compared to the expression of these costimulatory molecules in tissues taken from patients with aseptic loosening. They demonstrated the increased expression of costimulatory molecules in response to wear particles both at the bone implant interface and *in vitro*. These findings reinforce the hypothesis that macrophages have the potential to aid T-cell activation in response to metal or metal ions from orthopaedic implants, as well as to augment any T-cell mediated response.

Several studies have described perivascular lymphocytes in tissue membranes around failed MoM implants apparently not associated with infection, and the authors have interpreted this inflammation as an immunologic reaction against metal ions or metal particles associated with those articulations [131,139–142]. Polyzois *et al.* [143] reviewed the evidence of local and systemic toxicity of wear debris from total hip arthroplasty and found extensive evidence and experimental data supporting the fact that orthopaedic metals induce local immunological effects characterized by an unusual lymphocytic infiltration and cell-mediated hypersensitivity. Thomas *et al.* [144] reported the case of a patient who developed eczema and impaired wound-healing following the fixation of an ankle fracture with titanium-based implants. Histological analysis of the tissue around the implant demonstrated inflammation primarily with lymphocytes, and a contact allergy to nickel and cobalt was found in the absence of titanium hyper-reactivity, raising the question of a prior unknown nickel exposure as the source of the complications. Similarly, Gao *et al.* [145] reported a case of systemic dermatitis caused by Cr (serum Cr level of 61.9 μg/L) after total knee arthroplasty. Another feature reported as a metal-induced systemic T-lymphocyte-mediated hypersensitivity reaction is the formation of periprosthetic soft-tissue masses in patients with MoM devices [132,146]. A delayed T-lymphocyte-mediated self-perpetuating response can also create extensive tissue damage [122]. To date, the only means to predict/diagnose those individuals that will have an excessive immune response to metal exposure that may lead to premature implant failure are lymphocyte transformation test and patch testing (for skin reactions). However, they are not so useful in the evaluation of deep tissue metal allergy [122,145]. Although complications in metal-allergic patients appear to be generally rare [115], metal nanoparticles and high metal ion concentrations remain a concern as they could trigger early events leading to implant failure or a shorter implant lifespan in sensitized patients. Moreover, it remains unclear whether sensitization is a direct cause of implant loosening and failure, or if it is a consequence of particle loading due to device loosening.

3.3. Gene Expression

Over the past few years numerous investigations have been carried out to study the effects of different ions and particulate wear debris on the expression of an array of cytokine and toxicology related genes *in vitro*. TNF-α, IL-1, and IL-6 are cytokines that have been reported to play a central role in the induction of implant osteolysis [147,148]. Extensive work has been carried out on cytokine production by macrophages in response to wear debris. Sethi *et al.* [40] studied the macrophage response to cross-linked ultra-high molecular weight polyethylene (UHMWPE) and compared it to conventional UHMWPE as well as TiAlV and cobalt-chrome alloy (CoCr). At 24 and 48 h, macrophages cultured on TiAlV and CoCr alloy expressed higher levels of IL-1α, IL-1β, IL-6, and TNF-α than when grown on UHMWPE materials according to real time reverse transcription polymerase chain reaction (qRT-PCR) analysis. Jakobsen *et al.* [149] compared surfaces of as-cast and wrought CoCrMo alloy and TiAlV alloy when incubated with mouse macrophage J774A.1 cells and reported a significant increase in the levels of expression of TNF-α, IL-6, IL-1α, and β from cells incubated with alloys compared to non-stimulated cells. Garrigues *et al.* [150] used microarrays to investigate alterations in the phenotype of macrophages as they interact with UHMWPE and TiAlV alloy particulate wear debris. Their findings further validate the important roles of TNF-α, IL-1β, IL-1α, IL-6, MIP1α, and MIP1β in osteolysis. In a recent study, Posada *et al.* [43] examined the ability of the metal debris and Co ions to induce general toxicology-related gene expression of human monocyte-like U937 cells. In some experiments, they pre-treated the cells with Co ions prior to exposure to CoCr particles, in order to simulate the *in vivo* situation where a patient may receive a second MoM implant in either a bilateral or a revision procedure. Analysis of qRT-PCR results found significant up-regulation of inducible nitric oxide synthase (NOS2) and Bcl2-associated athanogene (BAG1) in Co pre-treated cells which were subsequently exposed to Co ions and debris. They showed that metal debris was more effective as an inducer of gene expression when cells had been pre-treated with Co ions. Overexpression of NOS2, which leads to an over production of NO, could have a predominant role in the inflammation and acidification of the peri-implant microenvironment, which in turn could exacerbate the corrosion of the nanoparticles. Co-expression of BAG1 and Bcl-2 has been shown to increase protection from cell death [151,152]. Consequently, up-regulation of BAG1 could be interpreted as part of a defence mechanism for delaying cell death in response to metal toxicity, particularly Co toxicity, in this case. Since the main gene expression fold changes were observed in cells pre-treated with Co ions, patients with a MoM implant undergoing revision surgery or receiving a second MoM device may potentially be at higher risk of implant failure.

As well as macrophages, other cell types have been reported as being involved in the biological response to implant wear debris. As a result, there are similar studies on monocytes, lymphocytes, osteocytes, and osteoblasts. For example, the effects of CoCr particles on osteocytes were tested by Kanaji *et al.* [148]. CoCr treatment of murine long bone osteocyte Y4 (MLO-Y4) osteocytes significantly up-regulated TNFα gene expression after 3 and 6 h and TNF-α protein production after 24 h, but down-regulated IL-6 gene expression after 6 h. MG-63 osteoblasts were treated by Vermes *et al.* [153] with titanium, titanium alloy, chromium orthophosphate, polyethylene, and polystyrene particles and they reported that each type of particle significantly suppressed procollagen alpha1[I] gene expression ($p < 0.05$), whereas other osteoblast-specific genes (osteonectin, osteocalcin, and alkaline phosphatase) did not show significant changes. The effect of particulate derivatives of nickel and cobalt alloys on the mRNA levels of chemokine receptors CCR1 and CCR2 on monocytes/macrophages from whole blood were analysed by Fujiyoshi and Hunt [154]. Although there were no significant differences in the level of CCR1 mRNA in monocytes/macrophages incubated with NiCr particulates, there was a down-regulation in the level of CCR2 in cells incubated with NiCr and CoCr particles. All these investigations indicate that wear debris and metal ions derived from MoM implants can cause an adverse tissue response by modulating gene expression on several types of cells, which suggests that osteolysis and subsequent aseptic loosening is the result of the concerted action of the different cell types.

Previous studies have stated that ions released from the wear debris could also affect gene expression. It has been reported that Cr^{+3} and Co^{+2} ions could induce damage to proteins in macrophage-like cells *in vitro*, probably through the formation of reactive oxygen species (ROS) [155,156]. U937 cells were exposed to Cr^{+6} and Co^{+2} ions by Tkaczyk *et al.* [157]. Cr^{+6} induced the protein expression of Mn-superoxide dismutase, Cu/Zn superoxide dismutase, catalase, glutathione peroxidase, and heme oxygenase-1 (HO-1) but had no effect on the expression of their mRNA, whereas Co^{+2} induced the expression of both protein and mRNA of HO-1 only. Co^{+2} had no effect on the expression of the other proteins. The overexpression of HO-1 has been suggested to play an important role in cellular protection against oxidant-mediated cell damage [158]. This suggests that the results from Tkaczyk *et al.* [157] show that Cr and Co ions cause oxidant-mediated cell damage. Type-I collagen gene expression was evaluated by Hallab *et al.* [159] after treating MG-63 cells with increasing concentrations (from 0.001 to 10 mM) of a variety of metal ions including Co and Cr. At toxic concentrations (1 mM), Co depressed osteoblast function by significantly decreasing the levels of type I collagen gene expression to 40% of control values. Queally *et al.* [160] showed that 10 ppm Co ions induce chemokine secretion in primary human osteoblasts by measuring the up-regulation of IL-8 and MCP-1 gene expression in osteoblasts stimulated with 0–10 ppm Co^{2+}. The level of expression of one of the principal proteinases capable of degrading native fibrillar collagens in the extracellular matrix, MMP-1, and its tissue inhibitors (TIMP-1) in U937 cells exposed to Co^{2+} and Cr^{3+} ions for 24 h, was determined by Luo *et al.* [161] who showed that these ions induce up-regulation in a dose-dependent manner. Their expression was studied to gain insight into the regulation of extracellular matrix degradation and tissue remodelling around hip prostheses. Altered expression of MMP-1 and TIMP-1 in the periprosthetic tissues has led to the hypothesis that their imbalance could contribute to the loosening of total hip prosthesis [162]. The findings from Luo *et al.* [161] suggest that Co and Cr ions can up-regulate the MMP-1 expression *in vivo*. These studies provide more evidence of potential gene expression modulation by wear debris and ions derived from MoM implants.

Receptor activator of nuclear factor-κB ligand (RANKL), its receptor, receptor activator of nuclear factor-κB (RANK), and its soluble inhibitor osteoprotegerin (OPG) are recognized as key regulators of osteoclast formation that regulate bone resorption in both health and disease [163]. Several studies have demonstrated the expression of mRNA encoding RANKL, OPG, and RANK in peri-implant tissues associated with osteolytic zones [164–167]. Jiang *et al.* [168] demonstrated a significantly elevated gene expression of RANKL in CoCr particle-challenged osteoblasts. Similarly, Pioletti and Kottelat [169] showed an increase of osteoblast gene expression for RANKL after exposure to Ti particles. Zijlstra *et al.* [170] determined the effects of Co and Cr ions on the expression of bone turnover regulatory proteins RANKL and OPG on human osteoblast-like cells. They found that the RANKL/OPG ratio increased after 72 h of exposure to 10 μg/L Co, 1 μg/L Cr, and higher, and at 1 μg/L Co + Cr and higher, indicating net bone loss. These findings are interesting since they seem to suggest that even in well-functioning MoM implants with systemic Co and Cr levels around 1 μg/L, local periprosthetic osteolytic reactions may take place. In a pilot study in our laboratory, gene expression of RANK, RANKL, and OPG in peripheral blood from six patients that had MoM hip implants for at least one year was investigated and correlated with the whole blood metal ion levels at the time of the analysis. There was a significant up-regulation of RANK and RANKL and significant down-regulation of OPG when compared to controls (no implant) (Figure 3). It has been suggested that the RANKL/OPG ratio is raised significantly in patients with severe osteolysis and that this imbalance is involved in bone resorption mechanisms [171]. Since OPG was down-regulated, patients had higher ratios (27.69 ± 10.53, mean $\pm$ SEM) when compared to controls (1 ± 0.15, mean $\pm$ SEM), suggesting an imbalance in the bone turnover system favouring bone resorption. However, a clear relationship between RANKL/OPG ratios and ion levels could not be established. Although changes in gene expression were identified, the lack of pre-surgery data made it impossible to determine whether the presence of the MoM implant was the cause of such changes.

All the studies mentioned in this section have been carried out in order to understand how wear debris affects the levels of expression of genes involved in osteolysis in tissues surrounding the joint implant. They suggest different mechanisms for transcriptional activation of the genes investigated, which could indicate that gene expression is modulated in a dose- and particle-dependent manner and as the result of several signals coming together.

Thus, the biological responses to metal wear debris are complex, involving regulation at different cellular and molecular levels to try and maintain intra- and extra-cellular homeostasis. When cells fail, they tip the balance towards inflammation and acidification. This acidification of the peri-implant microenvironment in turn enhances the corrosion of the nanoparticles and release of metal ions, which exacerbates the adverse reaction ultimately resulting in osteolysis and subsequent aseptic loosening.

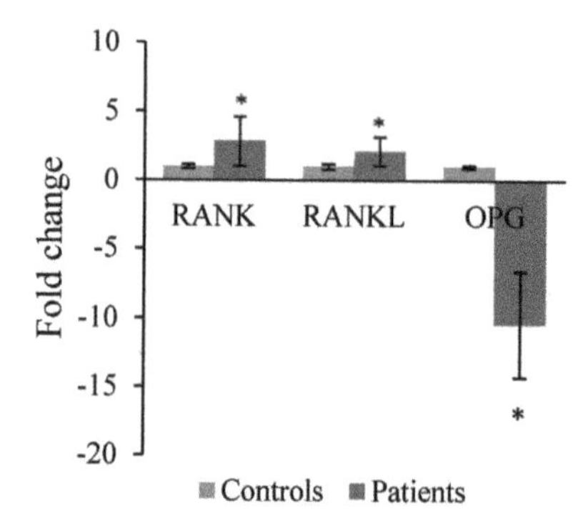

Figure 3. Fold variations of target genes in blood samples from controls and from patients with MoM hip implants. Results are mean values ± standard deviation (SD) (six biological samples with three technical replicates per gene assayed) expressed as the negative reciprocal. * Significantly different from control values by Analysis of variance (ANOVA) followed by Dunnet's comparison test ($p < 0.05$) (Posada *et al.*, unpublished data).

4. Discussion

Since the recall of MoM devices in 2010, the trends in bearing surface materials have changed, showing ceramic-on-polyethylene becoming more popular (Figure 4). The use of MoM has declined dramatically and the proportion of MoM resurfacing implants has decreased from a peak in 2006 to account for only 1.1% of implants in 2013 [117]. Although the recalls have taken most of the defective implant designs off the market, there are still tens of thousands of patients in the UK alone with these implants still *in situ*.

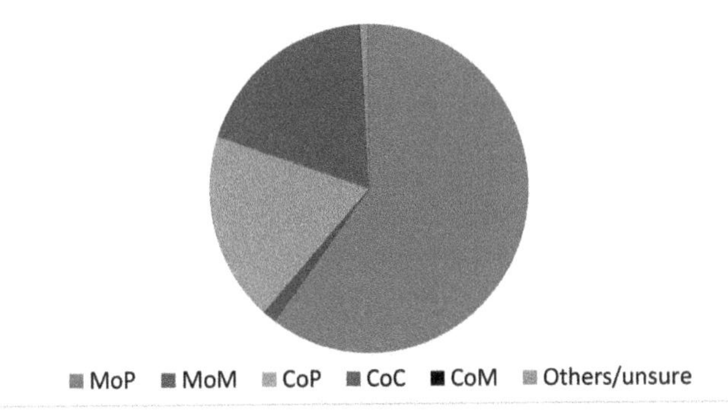

Figure 4. Number of primary hip replacements registered in 2013 by bearing surface. Metal-on-polyethylene (MoP): 45,944. Metal-on-metal (MoM, includes both total and resurfacing): 1014. Ceramic-on-polyethylene (CoP): 14,258. Ceramic-on-ceramic (CoC): 14,433. Ceramic-on-metal (CoM): 32. Others/unsure: 593 [3].

The Medicines and Healthcare Products Regulatory Agency (MHRA) issued information and advice about the follow-up of both symptomatic and asymptomatic patients implanted with MoM hip replacements, which include appropriate imaging (Metal Artifact Reduction Sequence (MARS) MRI/ultrasound), whole blood metal ion levels, and situations where revision may need to be considered. The MHRA has suggested that MARS MRI scans (or ultrasound scans) should carry more weight in decision-making than blood ion levels alone, and combined whole blood cobalt and chromium levels of greater than 7 ppb (7 μg/L), indicates potential for soft tissue reaction [172]. This decision is becoming more established among orthopaedic surgeons, but it does leave the quandary of the patients with circulating metal ion levels in several hundreds of μg/L and normal cross-sectional imaging. For these patients, many of whom have a well-functioning implant, it is difficult to understand whether or not the effects of particles and ions will ultimately impact their health locally or distally to their implant. It remains unclear whether these patients should be advised on a revision to avoid long-term systemic adverse effects on organs such as the heart and brain. It is worth noting that the 7 ppb threshold is a means of understanding how well the hip is performing *in vivo* [173], and revision is only considered if imaging is abnormal and/or blood metal ion levels rise [174]. Additionally, although this threshold is not based on soft tissue damage, levels of greater than 7 ppb are associated with significant soft-tissue reactions and failed MoM hips [63].

Although there has been widespread research on the various functions of different cytokines, questions concerning how inflammatory responses are triggered by wear particles remain largely unanswered. Specific organelles could play an important role in the cellular response triggered by wear particles. A growing body of evidence has suggested a role for endoplasmic reticulum (ER) stress in initiating inflammation, which is now thought to be fundamental to the pathogenesis of inflammatory diseases [175,176]. There is evidence suggesting that metal particles can cause increasing ER stress in various types of cells [177,178]. This presents the possibility that wear particles, produced around the prosthesis, have the potential to stimulate ER stress and thus may play a role in particle-induced osteolysis.

Chronic environmental exposure to some metal compounds, including arsenic, nickel, chromium, and cadmium, has been known to induce cancers and other diseases in exposed individuals [179]. While it has been shown that these metals disturb a vast array of cellular processes, such as redox state and various intracellular stress-signalling pathways, their ability to induce acute and/or chronic pathologies remains poorly understood. Sources of potential environmental exposure to these metals include occupational exposure and environmental contamination from industrial production [180]. Additionally, with all the evidence on particles and ions derived from metal orthopaedic implants, such replacements should be considered as an additional source of exposure. Emerging epidemiological studies show that the carcinogenic potential of some toxic metals may involve epigenetic changes, including silencing of DNA repair and tumor-suppressor genes [179]. The combinations of mechanisms, which confer long-term programming to genes and could bring about a change in gene function without changing gene sequence, are termed epigenetic [181]. As artificial articulations are being implanted in younger people, epigenetic studies could help assess how long-term exposure to metal debris and ions could bring about epigenetic changes altering gene expression which may have significant health-related consequences for these patients.

5. Conclusions

Despite the clinical success of hip replacements, this review has summarised some of the work addressing the still-present concern regarding the toxicity of metal particles and ions produced at the articulation site of a MoM implant. A vast literature on the effects of different kinds of debris as well as ions exists with many groups engaged in the discussion of mechanisms involved.

The potential dangers to patient health from CoCr alloy wear debris generation and the release of Co and Cr ions into circulation have been recognized. However, until the regulations for thoroughly testing the safety of medical devices such as hip replacements and other orthopaedic implants is more stringent, there is a very real possibility that 20 years from now we will be reading articles on the award of compensation for the next generation of articulations.

Acknowledgments: OMP is grateful for the financial support of an Overseas Research Scholarship and funds from the University of Strathclyde.

Author Contributions: The main author is Olga M. Posada, who wrote the first draft of the manuscript, and the revisions of this text to form the final draft were contributed by Rothwelle J. Tate, R. M. Dominic Meek, and M. Helen Grant. Dominic Meek provided clinical advice on the text.

Conflicts of Interest: The authors declare no conflict of interest.

References

1. Cobelli, N.; Scharf, B.; Crisi, G.M.; Hardin, J.; Santambrogio, L. Mediators of the inflammatory response to joint replacement devices. *Nat. Rev. Rheumatol.* **2011**, *7*, 600–608.
2. Australian Orthopaedic Association National Joint Replacement Registry. Annual Report 2014. Available online: https://aoanjrr.dmac.adelaide.edu.au/annual-reports-2014 (accessed on 20 January 2015).
3. National Joint Registry. 11th Annual Report. Available online: http://www.njrcentre.org.uk/njrcentre/Reports,PublicationsandMinutes/Annualreports/tabid/86/Default.aspx. (accessed on 20 January 2015).
4. Furnes, O.; Paxton, E.; Cafri, G.; Graves, S.; Bordini, B.; Comfort, T.; Rivas, M.C.; Banerjee, S.; Sedrakyan, A. Distributed analysis of hip implants using six national and regional registries: Comparing metal-on-metal with metal-on-highly cross-linked polyethylene bearings in cementless total hip arthroplasty in young patients. *J. Bone Joint Surg. Am.* **2014**, *96A*, 25–33.
5. Kanda, A.; Kaneko, K.; Obayashi, O.; Mogami, A. A 42-year-old patient presenting with femoral head migration after hemiarthroplasty performed 22 years earlier: A case report. *J. Med. Case Rep.* **2015**, *9*, 17.

6. Sedrakyan, A.; Romero, L.; Graves, S.; Davidson, D.; de Steiger, R.; Lewis, P.; Solomon, M.; Vial, R.; Lorimer, M. Survivorship of hip and knee implants in pediatric and young adult populations. *J. Bone Joint Surg. Am.* **2014**, *96A*, 73–78.
7. Eikmans, M.; Rekers, N.V.; Anholts, J.D.H.; Heidt, S.; Claas, F.H.J. Blood cell mRNAs and microRNAs: Optimized protocols for extraction and preservation. *Blood* **2013**, *121*, E81–E89.
8. Malchau, H.; Herberts, P.; Eisler, T.; Garellick, G.; Soderman, P. The swedish total hip replacement register. *J. Bone Joint Surg. Am.* **2002**, *84A*, 2–20.
9. Prokopovich, P. Interactions between mammalian cells and nano- or micro-sized wear particles: Physico-chemical views against biological approaches. *Adv. Colloid Interface Sci.* **2014**, *213*, 36–47.
10. Reito, A.; Puolakka, T.; Elo, P.; Pajamaki, J.; Eskelinen, A. High prevalence of adverse reactions to metal debris in small-headed ASR™ hips. *Clin. Orthop. Relat. Res.* **2013**, *471*, 2954–2961.
11. U.S. Food and Drug Administration (FDA). Medical devices. Available online: http://www.fda.gov/MedicalDevices/ProductsandMedicalProcedures/ImplantsandProsthetics/MetalonMetalHipImplants/ucm241770.htm (accessed on 20 January 2015).
12. Wynn-Jones, H.; Macnair, R.; Wimhurst, J.; Chirodian, N.; Derbyshire, B.; Toms, A.; Cahir, J. Silent soft tissue pathology is common with a modern metal-on-metal hip arthroplasty. *Acta Orthop.* **2011**, *82*, 301–307.
13. Langton, D.J.; Jameson, S.S.; Joyce, T.J.; Gandhi, J.N.; Sidaginamale, R.; Mereddy, P.; Lord, J.; Nargol, A.V.F. Accelerating failure rate of the asr total hip replacement. *J. Bone Joint Surg. Br.* **2011**, *93B*, 1011–1016.
14. Hayter, C.L.; Gold, S.L.; Koff, M.F.; Perino, G.; Nawabi, D.H.; Miller, T.T.; Potter, H.G. MRI findings in painful metal-on-metal hip arthroplasty. *Am. J. Roentgenol.* **2012**, *199*, 884–893.
15. Hart, A.J.; Hester, T.; Sinclair, K.; Powell, J.J.; Goodship, A.E.; Pele, L.; Fersht, N.L.; Skinner, J. The association between metal ions from hip resurfacing and reduced T-cell counts. *J. Bone Joint Surg. Br.* **2006**, *88*, 449–454.
16. Keegan, G.M.; Learmonth, I.D.; Case, C.P. Orthopaedic metals and their potential toxicity in the arthroplasty patient—Review of current knowledge and future strategies. *J. Bone Joint Surg. Br.* **2007**, *89B*, 567–573.
17. Moroni, A.; Savarino, L.; Hoque, M.; Cadossi, M.; Baldini, N. Do ion levels in hip resurfacing differ from metal-on-metal tha at midterm? *Clin. Orthop. Relat. Res.* **2011**, *469*, 180–187.
18. Hallab, N.J.; Jacobs, J.J. Biologic effects of implant debris. *Bull. NYU Hospital Joint Dis.* **2009**, *67*, 182–188.
19. Mathew, M.T.; Nagelli, C.; Pourzal, R.; Fischer, A.; Laurent, M.P.; Jacobs, J.J.; Wimmer, M.A. Tribolayer formation in a metal-on-metal (MoM) hip joint: An electrochemical investigation. *J. Mech. Behav. Biomed. Mater.* **2014**, *29*, 199–212.
20. Clayton, R.A.E.; Beggs, I.; Salter, D.M.; Grant, M.H.; Patton, J.T.; Porter, D.E. Inflammatory pseudotumor associated with femoral nerve palsy following metal-on-metal resurfacing of the hip. *J. Bone Joint Surg. Am.* **2008**, *90A*, 1988–1993.
21. Pandit, H.; Glyn-Jones, S.; McLardy-Smith, P.; Gundle, R.; Whitwell, D.; Gibbons, C.L.M.; Ostlere, S.; Athanasou, N.; Gill, H.S.; Murray, D.W. Pseudotumours associated with metal-on-metal hip resurfacings. *J. Bone Joint Surg. Br.* **2008**, *90B*, 847–851.
22. Dowson, D. Tribological principles in metal-on-metal hip joint design. *Proc. Inst. Mech. Eng. H J. Eng. Med.* **2006**, *220*, 161–171.
23. Yan, Y.; Neville, A.; Dowson, D. Biotribocorrosion—An appraisal of the time dependence of wear and corrosion interactions: I. The role of corrosion. *J. Phys. D Appl. Phys.* **2006**, *39*, 3200–3205.
24. Konttinen, Y.T.; Pajarinen, J. Surgery adverse reactions to metal-on-metal implants. *Nat. Rev. Rheumatol.* **2013**, *9*, 5–6.
25. Teoh, S.H. Fatigue of biomaterials: A review. *Int. J. Fatigue* **2000**, *22*, 825–837.
26. Ryu, J.J.; Shrotriya, P. Synergistic Mechanisms of Bio-Tribocorrosion in Medical Implants. In *Bio-Tribocorrosion in Biomaterials and Medical Implants*; Yan, Y., Ed.; Elsevier: Sawston, Cambridge, UK, 2013; pp. 25–44.
27. Hallab, N.J.; Jacobs, J.J. Orthopedic implant fretting corrosion. *Corros. Rev.* **2003**, *21*, 183–213.
28. Sargeant, A.; Goswami, T. Hip implants—Paper VI—Ion concentrations. *Mater. Des.* **2007**, *28*, 155–171.
29. Singh, R.; Dahotre, N.B. Corrosion degradation and prevention by surface modification of biometallic materials. *J. Mater. Sci. Mater. Med.* **2007**, *18*, 725–751.
30. Cadosch, D.; Chan, E.; Gautschi, O.P.; Filgueira, L. Metal is not inert: Role of metal ions released by biocorrosion in aseptic loosening-current concepts. *J. Biomed. Mater. Res. A* **2009**, *91A*, 1252–1262.

31. Doni, Z.; Alves, A.C.; Toptan, F.; Gomes, J.R.; Ramalho, A.; Buciumeanu, M.; Palaghian, L.; Silva, F.S. Dry sliding and tribocorrosion behaviour of hot pressed cocrmo biomedical alloy as compared with the cast CoCrMo and Ti6Al4V alloys. *Mater. Des.* **2013**, *52*, 47–57.
32. Zeng, Y.; Feng, W. Metal allergy in patients with total hip replacement: A review. *J. Int. Med. Res.* **2013**, *41*, 247–252.
33. Steinemann, S.G. Metal implants and surface reactions. *Inj. Int. J. Care Inj.* **1996**, *27*, 16–22.
34. Hanawa, T. Metal ion release from metal implants. *Mater. Sci. Eng. C Biomim. Supramol. Syst.* **2004**, *24*, 745–752.
35. Friesenbichler, J.; Maurer-Ertl, W.; Sadoghi, P.; Lovse, T.; Windhager, R.; Leithner, A. Serum metal ion levels after rotating-hinge knee arthroplasty: Comparison between a standard device and a megaprosthesis. *Int. Orthop.* **2012**, *36*, 539–544.
36. Penny, J.O.; Varmarken, J.E.; Ovesen, O.; Nielsen, C.; Overgaard, S. Metal ion levels and lymphocyte counts: ASR hip resurfacing prosthesis *vs.* standard THA 2-year results from a randomized study. *Acta Orthop.* **2013**, *84*, 130–137.
37. Gilbert, J.L.; Sivan, S.; Liu, Y.; Kocagoez, S.B.; Arnholt, C.M.; Kurtz, S.M. Direct *in vivo* inflammatory cell-induced corrosion of cocrmo alloy orthopedic implant surfaces. *J. Biomed. Mater. Res. A* **2015**, *103*, 211–223. [CrossRef] [PubMed]
38. Hallab, N.; Merritt, K.; Jacobs, J.J. Metal sensitivity in patients with orthopaedic implants. *J. Bone Joint Surg. Am.* **2001**, *83A*, 428–436.
39. Yang, J.; Black, J. Competitive-binding of chromium, cobalt and nickel to serum-proteins. *Biomaterials* **1994**, *15*, 262–268.
40. Sethi, R.K.; Neavyn, M.J.; Rubash, H.E.; Shanbhag, A.S. Macrophage response to cross-linked and conventional UHMWPE. *Biomaterials* **2003**, *24*, 2561–2573.
41. Rajamaki, K.; Nordstrom, T.; Nurmi, K.; Akerman, K.E.O.; Kovanen, P.T.; Oorni, K.; Eklund, K.K. Extracellular acidosis is a novel danger signal alerting innate immunity via the NLRP3 inflammasome. *J. Biol. Chem.* **2013**, *288*, 13410–13419.
42. Steen, K.H.; Steen, A.E.; Reeh, P.W. A dominant role of acid ph in inflammatory excitation and sensitization of nociceptors in rat skin, *in-vitro*. *J. Neurosci.* **1995**, *15*, 3982–3989.
43. Posada, O.M.; Gilmour, D.; Tate, R.J.; Grant, M.H. CoCr wear particles generated from cocr alloy metal-on-metal hip replacements, and cobalt ions stimulate apoptosis and expression of general toxicology-related genes in monocyte-like U937 cells. *Toxicol. Appl. Pharmacol.* **2014**, *281*, 125–135.
44. Geborek, P.; Saxne, T.; Pettersson, H.; Wollheim, F.A. Synovial-fluid acidosis correlates with radiological joint destruction in rheumatoid-arthritis knee joints. *J. Rheumatol.* **1989**, *16*, 468–472.
45. Mansson, B.; Geborek, P.; Saxne, T.; Bjornsson, S. Cytidine deaminase activity in synovial-fluid of patients with rheumatoid-arthritis—Relation to lactoferrin, acidosis, and cartilage proteoglycan release. *Ann. Rheum. Dis.* **1990**, *49*, 594–597.
46. Bagchi, D.; Stohs, S.J.; Downs, B.W.; Bagchi, M.; Preuss, H.G. Cytotoxicity and oxidative mechanisms of different forms of chromium. *Toxicology* **2002**, *180*, 5–22.
47. Gray, S.J.; Sterling, K. The tagging of red cells and plasma proteins with radioactive chromium. *J. Clin. Invest.* **1950**, *29*, 1604–1613.
48. Dillon, C.T.; Lay, P.A.; Bonin, A.M.; Cholewa, M.; Legge, G.J.F. Permeability, cytotoxicity, and genotoxicity of Cr(III) complexes and some Cr(V) analogues in V79 chinese hamster lung cells. *Chem. Res. Toxicol.* **2000**, *13*, 742–748.
49. Biedermann, K.A.; Landolph, J.R. Role of valence state and solubility of chromium compounds on induction of cytotoxicity, mutagenesis, and anchorage independence in diploid human fibroblasts. *Cancer Res.* **1990**, *50*, 7835–7842.
50. Tkaczyk, C.; Huk, O.L.; Mwale, F.; Antoniou, J.; Zukor, D.J.; Petit, A.; Tabrizian, M. The molecular structure of complexes formed by chromium or cobalt ions in simulated physiological fluids. *Biomaterials* **2009**, *30*, 460–467.
51. Raja, N.S.; Sankaranarayanan, K.; Dhathathreyan, A.; Nair, B.U. Interaction of chromium(III) complexes with model lipid bilayers: Implications on cellular uptake. *Biochim. Biophys. Acta* **2011**, *1808*, 332–340.
52. Fornsaglio, J.L.; O'Brien, T.J.; Patierno, S.R. Differential impact of ionic and coordinate covalent chromium (Cr)-DNA binding on DNA replication. *Mol. Cell Biochem.* **2005**, *279*, 149–155.

53. Shrivastava, H.Y.; Ravikumar, T.; Shanmugasundaram, N.; Babu, M.; Nair, B.U. Cytotoxicity studies of chromium(III) complexes on human dermal fibroblasts. *Free Radic. Biol. Med.* **2005**, *38*, 58–69.
54. Afolaranmi, G.A.; Tettey, J.; Meek, R.M.D.; Grant, M.H. Release of chromium from orthopaedic arthroplasties. *Open Orthop. J.* **2008**, *2*, 10–18.
55. Daniel, J.; Ziaee, H.; Pradhan, C.; Pynsent, P.B.; McMinn, D.J.W. Blood and urine metal ion levels in young and active patients after Birmingham hip resurfacing arthroplasty—Four-year results of a prospective longitudinal study. *J. Bone Joint Surg. Br.* **2007**, *89B*, 169–173.
56. Ziaee, H.; Daniel, J.; Datta, A.K.; Blunt, S.; McMinn, D.J.W. Transplacental transfer of cobalt and chromium in patients with metal-on-metal hip arthroplasty—A controlled study. *J. Bone Joint Surg. Br.* **2007**, *89B*, 301–305.
57. Antoniou, J.; Zukor, D.J.; Mwale, F.; Minarik, W.; Petit, A.; Huk, O.L. Metal ion levels in the blood of patients after hip resurfacing: A comparison between twenty-eight and thirty-six-millimeter-head metal-on-metal prostheses. *J. Bone Joint Surg. Am.* **2008**, *90A*, 142–148.
58. Wretenberg, P. Good function but very high concentrations of cobalt and chromium ions in blood 37 years after metal-on-metal total hip arthroplasy. *Med. Devices (Auckland, N.Z.)* **2008**, *1*, 31–32.
59. Hart, A.J.; Sabah, S.; Henckel, J.; Lewis, A.; Cobb, J.; Sampson, B.; Mitchell, A.; Skinner, J.A. The painful metal-on-metal hip resurfacing. *J. Bone Joint Surg. Br.* **2009**, *91B*, 738–744.
60. Hart, A.J.; Quinn, P.D.; Sampson, B.; Sandison, A.; Atkinson, K.D.; Skinner, J.A.; Powell, J.J.; Mosselmans, J.F.W. The chemical form of metallic debris in tissues surrounding metal-on-metal hips with unexplained failure. *Acta Biomater.* **2010**, *6*, 4439–4446.
61. Langton, D.J.; Jameson, S.S.; Joyce, T.J.; Hallab, N.J.; Natu, S.; Nargol, A.V.F. Early failure of metal-on-metal bearings in hip resurfacing and large-diameter total hip replacement a consequence of excess wear. *J. Bone Joint Surg. Br.* **2010**, *92B*, 38–46.
62. Davda, K.; Lali, F.V.; Sampson, B.; Skinner, J.A.; Hart, A.J. An analysis of metal ion levels in the joint fluid of symptomatic patients with metal-onmetal hip replacements. *J. Bone Joint Surg. Br.* **2011**, *93B*, 738–745.
63. Hart, A.J.; Sabah, S.A.; Bandi, A.S.; Maggiore, P.; Tarassoli, P.; Sampson, B.; Skinner, J.A. Sensitivity and specificity of blood cobalt and chromium metal ions for predicting failure of metal-on-metal hip replacement. *J. Bone Joint Surg. Br.* **2011**, *93B*, 1308–1313.
64. Malviya, A.; Ramaskandhan, J.R.; Bowman, R.; Kometa, S.; Hashmi, M.; Lingard, E.; Holland, J.P. What advantage is there to be gained using large modular metal-on-metal bearings in routine primary hip replacement? A preliminary report of a prospective randomised controlled trial. *J. Bone Joint Surg. Br.* **2011**, *93B*, 1602–1609.
65. Fritzsche, J.; Borisch, C.; Schaefer, C. Case report: High chromium and cobalt levels in a pregnant patient with bilateral metal-on-metal hip arthroplasties. *Clin. Orthop. Relat. Res.* **2012**, *470*, 2325–2331.
66. Matthies, A.K.; Skinner, J.A.; Osmani, H.; Henckel, J.; Hart, A.J. Pseudotumors are common in well-positioned low-wearing metal-on-metal hips. *Clin. Orthop. Relat. Res.* **2012**, *470*, 1895–1906.
67. Lass, R.; Grubl, A.; Kolb, A.; Stelzeneder, D.; Pilger, A.; Kubista, B.; Giurea, A.; Windhager, R. Comparison of synovial fluid, urine, and serum ion levels in metal-on-metal total hip arthroplasty at a minimum follow-up of 18 years. *J. Orthop. Res.* **2014**, *32*, 1234–1240.
68. Valko, M.; Morris, H.; Cronin, M.T.D. Metals, toxicity and oxidative stress. *Curr. Med. Chem.* **2005**, *12*, 1161–1208.
69. De Flora, S. Threshold mechanisms and site specificity in chromium(VI) carcinogenesis. *Carcinogenesis* **2000**, *21*, 533–541.
70. Merritt, K.; Brown, S.A. Release of hexavalent chromium from corrosion of stainless-steel and cobalt-chromium alloys. *J. Biomed. Mater. Res.* **1995**, *29*, 627–633.
71. MacDonald, S.J. Can a safe level for metal ions in patients with metal-on-metal total hip arthroplasties be determined? *J. Arthroplast.* **2004**, *19*, 71–77.
72. Nickens, K.P.; Patierno, S.R.; Ceryak, S. Chromium genotoxicity: A double-edged sword. *Chem. Biol. Interact.* **2010**, *188*, 276–288.
73. Codd, R.; Dillon, C.T.; Levina, A.; Lay, P.A. Studies on the genotoxicity of chromium: From the test tube to the cell. *Coord. Chem. Rev.* **2001**, *216*, 537–582.
74. Shettlemore, M.G.; Bundy, K.J. Examination of *in vivo* influences on bioluminescent microbial assessment of corrosion product toxicity. *Biomaterials* **2001**, *22*, 2215–2228.

75. De Smet, K.; de Haan, R.; Calistri, A.; Campbell, P.A.; Ebramzadeh, E.; Pattyn, C.; Gill, H.S. Metal ion measurement as a diagnostic tool to identify problems with metal-on-metal hip resurfacing. *J. Bone Joint Surg. Am.* **2008**, *90A*, 202–208.
76. Catalani, S.; Rizzetti, M.C.; Padovani, A.; Apostoli, P. Neurotoxicity of cobalt. *Hum. Exp. Toxicol.* **2012**, *31*, 421–437.
77. Kravenskaya, Y.V.; Fedirko, N.V. Mechanisms underlying interaction of zinc, lead, and cobalt with nonspecific permeability pores in the mitochondrial membranes. *Neurophysiology* **2011**, *43*, 163–172.
78. Bleackley, M.R.; MacGillivray, R.T.A. Transition metal homeostasis: From yeast to human disease. *Biometals* **2011**, *24*, 785–809.
79. Virginio, C.; Church, D.; North, R.A.; Surprenant, A. Effects of divalent cations, protons and calmidazolium at the rat P2X7 receptor. *Neuropharmacology* **1997**, *36*, 1285–1294.
80. Park, J.D.; Cherrington, N.J.; Klaassen, C.D. Intestinal absorption of cadmium is associated with divalent metal transporter 1 in rats. *Toxicol. Sci.* **2002**, *68*, 288–294.
81. Griffin, K.P.; Ward, D.T.; Liu, W.; Stewart, G.; Morris, I.D.; Smith, C.P. Differential expression of divalent metal transporter DMT1 (Slc11a2) in the spermatogenic epithelium of the developing and adult rat testis. *Am. J. Physiol. Cell Physiol.* **2005**, *288*, C176–C184.
82. Forbes, J.R.; Gros, P. Iron, manganese, and cobalt transport by Nramp1 (Slc11a1) and Nramp2 (Slc11a2) expressed at the plasma membrane. *Blood* **2003**, *102*, 1884–1892.
83. De Boeck, M.; Kirsch-Volders, M.; Lison, D. Cobalt and antimony: Genotoxicity and carcinogenicity. *Mutat. Res.* **2003**, *533*, 135–152.
84. Catelas, I.; Wimmer, M.A. New insights into wear and biological effects of metal-on-metal bearings. *J. Bone Joint Surg. Am.* **2011**, *93A*, 76–83.
85. Lucarelli, M.; Gatti, A.M.; Savarino, G.; Quattroni, P.; Martinelli, L.; Monari, E.; Boraschi, D. Innate defence functions of macrophages can be biased by nano-sized ceramic and metallic particles. *Eur. Cytokine Netw.* **2004**, *15*, 339–346.
86. Papageorgiou, I.; Brown, C.; Schins, R.; Singh, S.; Newson, R.; Davis, S.; Fisher, J.; Ingham, E.; Case, C.P. The effect of nano- and micron-sized particles of cobalt-chromium alloy on human fibroblasts *in vitro*. *Biomaterials* **2007**, *28*, 2946–2958.
87. Dalal, A.; Pawar, V.; McAllister, K.; Weaver, C.; Hallab, N.J. Orthopedic implant cobalt-alloy particles produce greater toxicity and inflammatory cytokines than titanium alloy and zirconium alloy-based particles *in vitro*, in human osteoblasts, fibroblasts, and macrophages. *J. Biomed. Mater. Res. A* **2012**, *100A*, 2147–2158.
88. Posada, O.M.; Tate, R.J.; Grant, M.H. Effects of CoCr metal wear debris generated from metal-on-metal hip implants and co ions on human monocyte-like U937 cells. *Toxicol. Vitro* **2014**, *29*, 271–280.
89. Simonsen, L.O.; Harbak, H.; Bennekou, P. Cobalt metabolism and toxicology-A brief update. *Sci. Total Environ.* **2012**, *432*, 210–215.
90. VanOs, R.; Lildhar, L.L.; Lehoux, E.A.; Beaule, P.E.; Catelas, I. *In vitro* macrophage response to nanometer-size chromium oxide particles. *J. Biomed. Mater. Res. B Appl. Biomater.* **2014**, *102*, 149–159.
91. Tsaousi, A.; Jones, E.; Case, C.P. The *in vitro* genotoxicity of orthopaedic ceramic (Al_2O_3) and metal (CoCr alloy) particles. *Mutat. Res.* **2010**, *697*, 1–9.
92. Araya, J.; Maruyama, M.; Inoue, A.; Fujita, T.; Kawahara, J.; Sassa, K.; Hayashi, R.; Kawagishi, Y.; Yamashita, N.; Sugiyama, E.; *et al.* Inhibition of proteasome activity is involved in cobalt-induced apoptosis of human alveolar macrophages. *Am. J. Physiol. Lung Cell. Mol. Physiol.* **2002**, *283*, L849–L858.
93. Zou, W.G.; Yan, M.D.; Xu, W.J.; Huo, H.R.; Sun, L.Y.; Zheng, Z.C.; Liu, X.Y. Cobalt chloride induces PC12 cells apoptosis through reactive oxygen species and accompanied by AP-1 activation. *J. Neurosci. Res.* **2001**, *64*, 646–653.
94. Akbar, M.; Brewer, J.M.; Grant, M.H. Effect of chromium and cobalt ions on primary human lymphocytes *in vitro*. *J. Immunotoxicol.* **2011**, *8*, 140–149.
95. Catelas, I.; Petit, A.; Vali, H.; Fragiskatos, C.; Meilleur, R.; Zukor, D.J.; Antoniou, J.; Huk, O.L. Quantitative analysis of macrophage apoptosis *vs.* Necrosis induced by cobalt and chromium ions *in vitro*. *Biomaterials* **2005**, *26*, 2441–2453.
96. Catelas, I.; Petit, A.; Zukor, D.J.; Huk, O.L. Cytotoxic and apoptotic effects of cobalt and chromium ions on J774 macrophages—Implication of caspase-3 in the apoptotic pathway. *J. Mater. Sci. Mater. Med.* **2001**, *12*, 949–953.

97. Catelas, I.; Petit, A.; Zukor, D.J.; Antoniou, J.; Huk, O.L. TNF-alpha secretion and macrophage mortality induced by cobalt and chromium ions *in vitro*—Qualitative analysis of apoptosis. *Biomaterials* **2003**, *24*, 383–391.
98. Caicedo, M.; Jacobs, J.J.; Reddy, A.; Hallab, N.J. Analysis of metal ion-induced DNA damage, apoptosis, and necrosis in human (Jurkat) T-cells demonstrates Ni^{2+}, and V^{3+} are more toxic than other metals: Al^{3+}, Be^{2+}, Co^{2+}, Cr^{3+}, Cu^{2+}, Fe^{3+}, Mo^{5+}, Nb^{5+}, Zr^{2+}. *J. Biomed. Mater. Res. A* **2008**, *86A*, 905–913.
99. Huber, M.; Reinisch, G.; Zenz, P.; Zweymueller, K.; Lintner, F. Postmortem study of femoral osteolysis associated with metal-on-metal articulation in total hip replacement an analysis of nine cases. *J. Bone Joint Surg. Am.* **2010**, *92A*, 1720–1731.
100. Langton, D.J.; Joyce, T.J.; Jameson, S.S.; Lord, J.; Van Orsouw, M.; Holland, J.P.; Nargol, A.V.F.; de Smet, K.A. Adverse reaction to metal debris following hip resurfacing the influence of component type, orientation and volumetric wear. *J. Bone Joint Surg. Br.* **2011**, *93B*, 164–171.
101. Case, C.P.; Langkamer, V.G.; James, C.; Palmer, M.R.; Kemp, A.J.; Heap, P.F.; Solomon, L. Widespread dissemination of metal debris from implants. *J. Bone Joint Surg. Br.* **1994**, *76*, 701–712.
102. Tower, S.S. Arthroprosthetic cobaltism: Neurological and cardiac manifestations in two patients with metal-on-metal arthroplasty a case report. *J. Bone Joint Surg. Am.* **2010**, *92*, 2847–2851.
103. Tower, S.S. Metal on metal hip implants arthroprosthetic cobaltism associated with metal on metal hip implants. *Br. Med. J.* **2012**, *344*, e430.
104. Oldenburg, M.; Wegner, R.; Baur, X. Severe cobalt intoxication due to prosthesis wear in repeated total hip arthroplasty. *J. Arthroplast.* **2009**, *24*, 825.e815–825.e820.
105. Steens, W.; von Foerster, G.; Katzer, A. Severe cobalt poisoning with loss of sight after ceramic-metal pairing in a hip—A case report. *Acta Orthop.* **2006**, *77*, 830–832.
106. Ikeda, T.; Takahashi, K.; Kabata, T.; Sakagoshi, D.; Tomita, K.; Yamada, M. Polyneuropathy caused by cobalt-chromium metallosis after total hip replacement. *Muscle Nerve* **2010**, *42*, 140–143.
107. Machado, C.; Appelbe, A.; Wood, R. Arthroprosthetic cobaltism and cardiomyopathy. *Heart Lung Circ.* **2012**, *21*, 759–760.
108. Pelclova, D.; Sklensky, M.; Janicek, P.; Lach, K. Severe cobalt intoxication following hip replacement revision: Clinical features and outcome. *Clin. Toxicol.* **2012**, *50*, 262–265.
109. Rizzetti, M.C.; Liberini, P.; Zarattini, G.; Catalani, S.; Pazzaglia, U.; Apostoli, P.; Padovani, A. Loss of sight and sound. Could it be the hip? *Lancet* **2009**, *373*, 1052.
110. Devlin, J.J.; Pomerleau, A.C.; Brent, J.; Morgan, B.W.; Deitchman, S.; Schwartz, M. Clinical features, testing, and management of patients with suspected prosthetic hip-associated cobalt toxicity: A systematic review of cases. *J. Med. Toxicol.* **2013**, *9*, 405–415.
111. Bradberry, S.M.; Wilkinson, J.M.; Ferner, R.E. Systemic toxicity related to metal hip prostheses. *Clin. Toxicol.* **2014**, *52*, 837–847.
112. Clark, M.J.; Prentice, J.R.; Hoggard, N.; Paley, M.N.; Hadjivassiliou, M.; Wilkinson, J.M. Brain structure and function in patients after metal-on-metal hip resurfacing. *Am. J. Neuroradiol.* **2014**, *35*, 1753–1758.
113. Makela, K.T.; Visuri, T.; Pulkkinen, P.; Eskelinen, A.; Remes, V.; Virolainen, P.; Junnila, M.; Pukkala, E. Risk of cancer with metal-on-metal hip replacements: Population based study. *Br. Med. J.* **2012**, *345*, e4646.
114. Christian, W.V.; Oliver, L.D.; Paustenbach, D.J.; Kreider, M.L.; Finley, B.L. Toxicology-based cancer causation analysis of cocr-containing hip implants: A quantitative assessment of genotoxicity and tumorigenicity studies. *J. Appl. Toxicol.* **2014**, *34*, 939–967.
115. Frigerio, E.; Pigatto, P.D.; Guzzi, G.; Altomare, G. Metal sensitivity in patients with orthopaedic implants: A prospective study. *Contact Dermat.* **2011**, *64*, 273–279.
116. Romesburg, J.W.; Wasserman, P.L.; Schoppe, C.H. Metallosis and metal-induced synovitis following total knee arthroplasty: Review of radiographic and CT findings. *J. Radiol. Case Rep.* **2010**, *4*, 7–17.
117. Desrochers, J.; Amrein, M.W.; Matyas, J.R. Microscale surface friction of articular cartilage in early osteoarthritis. *J. Mech. Behav. Biomed. Mater.* **2013**, *25*, 11–22.
118. Anderson, J.M.; Rodriguez, A.; Chang, D.T. Foreign body reaction to biomaterials. *Semin. Immunol.* **2008**, *20*, 86–100.
119. Kaufman, A.M.; Alabre, C.I.; Rubash, H.E.; Shanbhag, A.S. Human macrophage response to uhmwpe, tialv, cocr, and alumina particles: Analysis of multiple cytokines using protein arrays. *J. Biomed. Mater. Res. A* **2008**, *84A*, 464–474.

120. Cachinho, S.C.P.; Pu, F.R.; Hunt, J.A. Cytokine secretion from human peripheral blood mononuclear cells cultured *in vitro* with metal particles. *J. Biomed. Mater. Res. A* **2013**, *101A*, 1201–1209.
121. Devitt, B.M.; Queally, J.M.; Vioreanu, M.; Butler, J.S.; Murray, D.; Doran, P.P.; O'Byrne, J.M. Cobalt ions induce chemokine secretion in a variety of systemic cell lines. *Acta Orthop.* **2010**, *81*, 756–764.
122. Landgraeber, S.; Jaeger, M.; Jacobs, J.J.; Hallab, N.J. The pathology of orthopedic implant failure is mediated by innate immune system cytokines. *Mediat. Inflamm.* **2014**, *2014*, 185150.
123. Valladares, R.D.; Nich, C.; Zwingenberger, S.; Li, C.; Swank, K.R.; Gibon, E.; Rao, A.J.; Yao, Z.; Goodman, S.B. Toll-like receptors-2 and 4 are overexpressed in an experimental model of particle-induced osteolysis. *J. Biomed. Mater. Res. A* **2014**, *102*, 3004–3011.
124. Tyson-Capper, A.J.; Lawrence, H.; Holland, J.P.; Deehan, D.J.; Kirby, J.A. Metal-on-metal hips: Cobalt can induce an endotoxin-like response. *Ann. Rheum. Dis.* **2013**, *72*, 460–461.
125. Potnis, P.A.; Dutta, D.K.; Wood, S.C. Toll-like receptor 4 signaling pathway mediates proinflammatory immune response to cobalt-alloy particles. *Cell. Immunol.* **2013**, *282*, 53–65.
126. Werling, D.; Jungi, T.W. Toll-like receptors linking innate and adaptive immune response. *Vet. Immunol. Immunopathol.* **2003**, *91*, 1–12.
127. Piccinini, A.M.; Midwood, K.S. Dampening inflammation by modulating TLR signalling. *Mediat. Inflamm.* **2010**, *2010*, 1.
128. Minang, J.T.; Arestrom, I.; Troye-Blomberg, M.; Lundeberg, L.; Ahlborg, N. Nickel, cobalt, chromium, palladium and gold induce a mixed Th1- and Th2-type cytokine response *in vitro* in subjects with contact allergy to the respective metals. *Clin. Exp. Immunol.* **2006**, *146*, 417–426.
129. Hegewald, J.; Uter, W.; Pfahlberg, A.; Geier, J.; Schnuch, A.; IVDK. A multifactorial analysis of concurrent patch-test reactions to nickel, cobalt, and chromate. *Allergy* **2005**, *60*, 372–378.
130. Fors, R.; Stenberg, B.; Stenlund, H.; Persson, M. Nickel allergy in relation to piercing and orthodontic appliances—A population study. *Contact Dermatitis* **2012**, *67*, 342–350.
131. Willert, H.G.; Buchhorn, G.H.; Fayyazi, A.; Flury, R.; Windler, M.; Koster, G.; Lohmann, C.H. Metal-on-metal bearings and hypersensitivity in patients with artificial hip joints—A clinical and histomorphological study. *J. Bone Joint Surg. Am.* **2005**, *87A*, 28–36.
132. Boardman, D.R.; Middleton, F.R.; Kavanagh, T.G. A benign psoas mass following metal-on-metal resurfacing of the hip. *J. Bone Joint Surg. Br.* **2006**, *88B*, 402–404.
133. Moulon, C.; Vollmer, J.; Weltzien, H.U. Characterization of processing requirements acid metal cross-reactivities in T cell clones from patients with allergic contact dermatitis to nickel. *Eur. J. Immunol.* **1995**, *25*, 3308–3315.
134. Villanueva, M.B.G.; Koizumi, S.; Jonai, H. Cytokine production by human peripheral blood mononuclear cells after exposure to heavy metals. *J. Health Sci.* **2000**, *46*, 358–362.
135. Jiranek, W.A.; Machado, M.; Jasty, M.; Jevsevar, D.; Wolfe, H.J.; Goldring, S.R.; Goldberg, M.J.; Harris, W.H. Production of cytokines around loosened cemented acetabular components—Analysis with immunohistochemical techniques and *in-situ* hybridization. *J. Bone Joint Surg. Am.* **1993**, *75A*, 863–879.
136. Goodman, S.B.; Huie, P.; Song, Y.; Schurman, D.; Maloney, W.; Woolson, S.; Sibley, R. Cellular profile and cytokine production at prosthetic interfaces—Study of tissues retrieved from revised hip and knee replacements. *J. Bone Joint Surg. Br.* **1998**, *80B*, 531–539.
137. Voronov, I.; Santerre, J.P.; Hinek, A.; Callahan, J.W.; Sandhu, J.; Boynton, E.L. Macrophage phagocytosis of polyethylene particulate *in vitro*. *J. Biomed. Mater. Res.* **1998**, *39*, 40–51.
138. Bainbridge, J.A.; Revell, P.A.; Al-Saffar, N. Costimulatory molecule expression following exposure to orthopaedic implants wear debris. *J. Biomed. Mater. Res.* **2001**, *54*, 328–334.
139. Bohler, M.; Kanz, F.; Schwarz, B.; Steffan, I.; Walter, A.; Plenk, H.; Knahr, K. Adverse tissue reactions to wear particles form co-alloy articulations, increased by alumina-blasting particle contamination from cementless Ti-based total hip implants—A report of seven revisions with early failure. *J. Bone Joint Surg. Br.* **2002**, *84B*, 128–136.
140. Campbell, P.; Ebramzadeh, E.; Nelson, S.; Takamura, K.; de Smet, K.; Amstutz, H.C. Histological features of pseudotumor-like tissues from metal-on-metal hips. *Clin. Orthop. Relat. Res.* **2010**, *468*, 2321–2327.
141. Davies, A.P.; Willert, H.G.; Campbell, P.A.; Learmonth, I.D.; Case, C.P. An unusual lymphocytic perivascular infiltration in tissues around contemporary metal-on-metal joint replacements. *J. Bone Joint Surg. Am.* **2005**, *87A*, 18–27.

142. Korovessis, P.; Petsinis, G.; Repanti, M.; Repantis, T. Metallosis after contemporary metal-on-metal total hip arthroplasty—Five to nine-year follow-up. *J. Bone Joint Surg. Am.* **2006**, *88A*, 1183–1191.
143. Polyzois, I.; Nikolopoulos, D.; Michos, I.; Patsouris, E.; Theocharis, S. Local and systemic toxicity of nanoscale debris particles in total hip arthroplasty. *J. Appl. Toxicol.* **2012**, *32*, 255–269.
144. Thomas, P.; Thomas, M.; Summer, B.; Dietrich, K.; Zauzig, M.; Steinhauser, E.; Krenn, V.; Arnholdt, H.; Flaig, M.J. Impaired wound-healing, local eczema, and chronic inflammation following titanium osteosynthesis in a nickel and cobalt-allergic patient: A case report and review of the literature. *J. Bone Joint Surg. Am.* **2011**, *93*, e61.
145. Gao, X.; He, R.-X.; Yan, S.-G.; Wu, L.-D. Dermatitis associated with chromium following total knee arthroplasty. *J. Arthroplast.* **2011**, *26*. [CrossRef]
146. Kwon, Y.M.; Thomas, P.; Summer, B.; Pandit, H.; Taylor, A.; Beard, D.; Murray, D.W.; Gill, H.S. Lymphocyte proliferation responses in patients with pseudotumors following metal-on-metal hip resurfacing arthroplasty. *J. Orthop. Res.* **2010**, *28*, 444–450.
147. Masui, T.; Sakano, S.; Hasegawa, Y.; Warashina, H.; Ishiguro, N. Expression of inflammatory cytokines, rankl and opg induced by titanium, cobalt-chromium and polyethylene particles. *Biomaterials* **2005**, *26*, 1695–1702.
148. Kanaji, A.; Caicedo, M.S.; Virdi, A.S.; Sumner, D.R.; Hallab, N.J.; Sena, K. Co-Cr-Mo alloy particles induce tumor necrosis factor alpha production in MLO-Y4 osteocytes: A role for osteocytes in particle-induced inflammation. *Bone* **2009**, *45*, 528–533.
149. Jakobsen, S.S.; Larsen, A.; Stoltenberg, M.; Bruun, J.M.; Soballe, K. Effects of as-cast and wrought cobalt-chrome-molybdenum and titanium-aluminium-vanadium alloys on cytokine gene expression and protein secretion in J774a.1 macrophages. *Eur. Cells Mater.* **2007**, *14*, 45–54.
150. Garrigues, G.E.; Cho, D.R.; Rubash, H.E.; Goldring, S.R.; Herndon, J.H.; Shanbhag, A.S. Gene expression clustering using self-organizing maps: Analysis of the macrophage response to particulate biomaterials. *Biomaterials* **2005**, *26*, 2933–2945.
151. Takayama, S.; Sato, T.; Krajewski, S.; Kochel, K.; Irie, S.; Millan, J.A.; Reed, J.C. Cloning and functional-analysis of BAG-1—A novel Bcl-2-binding protein with anti-cell death activity. *Cell* **1995**, *80*, 279–284.
152. Terada, S.; Komatsu, T.; Fujita, T.; Terakawa, A.; Nagamune, T.; Takayama, S.; Reed, J.C.; Suzuki, E. Co-expression of Bcl-2 and BAG-1, apoptosis suppressing genes, prolonged viable culture period of hybridoma and enhanced antibody production. *Cytotechnology* **1999**, *31*, 143–151.
153. Vermes, C.; Chandrasekaran, R.; Jacobs, J.J.; Galante, J.O.; Roebuck, K.A.; Glant, T.T. The effects of particulate wear debris, cytokines, and growth factors on the functions of MG-63 osteoblasts. *J. Bone Joint Surg. Am.* **2001**, *83A*, 201–211.
154. Fujiyoshi, K.; Hunt, J.A. The effect of particulate material on the regulation of chemokine receptor expression in leukocytes. *Biomaterials* **2006**, *27*, 3888–3896.
155. Petit, A.; Mwale, F.; Tkaczyk, C.; Antoniou, J.; Zukor, D.J.; Huk, O.L. Induction of protein oxidation by cobalt and chromium ions in human U937 macrophages. *Biomaterials* **2005**, *26*, 4416–4422.
156. Petit, A.; Mwale, F.; Tkaczyk, C.; Antoniou, J.; Zukor, D.J.; Huk, O.L. Cobalt and chromium ions induce nitration of proteins in human U937 macrophages *in vitro*. *J. Biomed. Mater. Res. A* **2006**, *79A*, 599–605.
157. Tkaczyk, C.; Huk, O.L.; Mwale, F.; Antoniou, J.; Zukor, D.J.; Petit, A.; Tabrizian, M. Effect of chromium and cobalt ions on the expression of antioxidant enzymes in human U937 macrophage-like cells. *J. Biomed. Mater. Res. A* **2010**, *94A*, 419–425.
158. Rothfuss, A.; Speit, G. Overexpression of heme oxygenase-1 (HO-1) in V79 cells results in increased resistance to hyperbaric oxygen (HBO)-induced DNA damage. *Environ. Mol. Mutagen.* **2002**, *40*, 258–265.
159. Hallab, N.J.; Vermes, C.; Messina, C.; Roebuck, K.A.; Glant, T.T.; Jacobs, J.J. Concentration- and composition-dependent effects of metal ions on human MG-63 osteoblasts. *J. Biomed. Mater. Res.* **2002**, *60*, 420–433.
160. Queally, J.M.; Devitt, B.M.; Butler, J.S.; Malizia, A.P.; Murray, D.; Doran, P.P.; O'Byrne, J.M. Cobalt ions induce chemokine secretion in primary human osteoblasts. *J. Orthop. Res.* **2009**, *27*, 855–864.
161. Luo, L.; Petit, A.; Antoniou, J.; Zukor, D.J.; Huk, O.L.; Liu, R.C.W.; Winnik, F.M.; Mwale, F. Effect of cobalt and chromium ions on MMP-1 TIMP-1, and TNF-alpha gene expression in human U937 macrophages: A role for tyrosine kinases. *Biomaterials* **2005**, *26*, 5587–5593.

162. Takagi, M. Neutral proteinases and their inhibitors in the loosening of total hip prostheses. *Acta Orthop. Scand.* **1996**, *67*, 1–29.
163. Crotti, T.N.; Smith, M.D.; Findlay, D.M.; Zreiqat, H.; Ahern, M.J.; Weedon, H.; Hatzinikolous, G.; Capone, M.; Holding, C.; Haynes, D.R. Factors regulating osteoclast formation in human tissues adjacent to peri-implant bone loss: Expression of receptor activator NF kappaB, RANK ligand and osteoprotegerin. *Biomaterials* **2004**, *25*, 565–573.
164. Haynes, D.R.; Crotti, T.N.; Loric, M.; Bain, G.I.; Atkins, G.J.; Findlay, D.M. Osteoprotegerin and receptor activator of nuclear factor kappaB ligand (RANKL) regulate osteoclast formation by cells in the human rheumatoid arthritic joint. *Rheumatology* **2001**, *40*, 623–630.
165. Holding, C.A.; Findlay, D.M.; Stamenkov, R.; Neale, S.D.; Lucas, H.; Dharmapatni, A.; Callary, S.A.; Shrestha, K.R.; Atkins, G.J.; Howie, D.W.; *et al.* The correlation of RANK, RANKL and TNF alpha expression with bone loss volume and polyethylene wear debris around hip implants. *Biomaterials* **2006**, *27*, 5212–5219.
166. Mao, X.; Pan, X.Y.; Zhao, S.; Peng, X.C.; Cheng, T.; Zhang, X.L. Protection against titanium particle-induced inflammatory osteolysis by the proteasome inhibitor bortezomib *in vivo*. *Inflammation* **2012**, *35*, 1378–1391.
167. Chen, D.S.; Zhang, X.L.; Guo, Y.Y.; Shi, S.F.; Mao, X.; Pan, X.Y.; Cheng, T. MMP-9 inhibition suppresses wear debris-induced inflammatory osteolysis through downregulation of RANK/RANKL in a murine osteolysis model. *Int. J. Mol. Med.* **2012**, *30*, 1417–1423.
168. Jiang, Y.; Jia, T.; Gong, W.; Wooley, P.H.; Yang, S.-Y. Effects of Ti, PMMA, UHMWPE, and Co-Cr wear particles on differentiation and functions of bone marrow stromal cells. *J. Biomed. Mater. Res. A* **2013**, *101*, 2817–2825.
169. Pioletti, D.P.; Kottelat, A. The influence of wear particles in the expression of osteoclastogenesis factors by osteoblasts. *Biomaterials* **2004**, *25*, 5803–5808.
170. Zijlstra, W.P.; Bulstra, S.K.; van Raay, J.; van Leeuwen, B.M.; Kuijer, R. Cobalt and chromium ions reduce human osteoblast-like cell activity *in vitro*, reduce the OPG to RANKL ratio, and induce oxidative stress. *J. Orthop. Res.* **2012**, *30*, 740–747.
171. Perez-Sayans, M.; Manuel Somoza-Marin, J.; Barros-Angueira, F.; Gandara Rey, J.M.; Garcia-Garcia, A. RANK/RANKL/OPG role in distraction osteogenesis. *Oral Surg. Oral Med. Oral Pathol. Oral Radiol. Endodontol.* **2010**, *109*, 679–686.
172. Medicines and Healthcare Products Regulatory agency (MHRA). Medical device alert: All metal-on-metal (MoM) hip replacements (MDA/2012/036). Available online: http://www.mhra.gov.uk/Publications/Safetywarnings/MedicalDeviceAlerts/CON155761. (accessed on 20 January 2015).
173. Sampson, B.; Hart, A. Clinical usefulness of blood metal measurements to assess the failure of metal-on-metal hip implants. *Ann. Clin. Biochem.* **2012**, *49*, 118–131.
174. Anderson, J.M. *In vitro* and *in vivo* monocyte, macrophage, foreign body giant cell, and lymphocyte interactions with biomaterials. In *Biological Interactions on Material Surfaces*; Springer US: New York, NY, USA, 2009; pp. 225–244.
175. Zhang, K.; Kaufman, R.J. From endoplasmic-reticulum stress to the inflammatory response. *Nature* **2008**, *454*, 455–462.
176. Hansson, G.K.; Libby, P. The immune response in atherosclerosis: A double-edged sword. *Nat. Rev. Immunol.* **2006**, *6*, 508–519.
177. Tsai, Y.-Y.; Huang, Y.-H.; Chao, Y.-L.; Hu, K.-Y.; Chin, L.-T.; Chou, S.-H.; Hour, A.-L.; Yao, Y.-D.; Tu, C.-S.; Liang, Y.-J.; *et al.* Identification of the nanogold particle-induced endoplasmic reticulum stress by omic techniques and systems biology analysis. *ACS Nano* **2011**, *5*, 9354–9369.
178. Zhang, R.; Piao, M.J.; Kim, K.C.; Kim, A.D.; Choi, J.-Y.; Choi, J.; Hyun, J.W. Endoplasmic reticulum stress signaling is involved in silver nanoparticles-induced apoptosis. *Int. J. Biochem. Cell Biol.* **2012**, *44*, 224–232.
179. Martinez-Zamudio, R.; Ha, H.C. Environmental epigenetics in metal exposure. *Epigenetics* **2011**, *6*, 820–827.
180. Salnikow, K.; Zhitkovich, A. Genetic and epigenetic mechanisms in metal carcinogenesis and cocarcinogenesis: Nickel, arsenic, and chromium. *Chem. Res. Toxicol.* **2008**, *21*, 28–44.
181. Stoccoro, A.; Karlsson, H.L.; Coppede, F.; Migliore, L. Epigenetic effects of nano-sized materials. *Toxicology* **2013**, *313*, 3–14. [CrossRef] [PubMed]

Article

Ultra-High Molecular Weight Polyethylene Reinforced with Multiwall Carbon Nanotubes: *In Vitro* Biocompatibility Study Using Macrophage-Like Cells

Nayeli Camacho [1,*], Stephen W. Stafford [1], Kristine M. Garza [2], Raquel Suro [3] and Kristina I. Barron [3]

1 Metallurgical and Materials Engineering Department, University of Texas at El Paso, El Paso, TX 79968, USA; stafford@utep.edu
2 Border Biomedical Research Center, University of Texas at El Paso, El Paso, TX 79968, USA; kgarza@utep.edu
3 Biological Science Department, University of Texas at El Paso, El Paso, TX 79968, USA; raquelmsuro@gmail.com (R.S.); kibarron@miners.utep.edu (K.I.B.)
* Author to whom correspondence should be addressed; ncamacho@miners.utep.edu; Tel.: +1-915-747-6930; Fax: +1-915-747-8036.

Academic Editor: James E. Krzanowski
Received: 1 June 2015; Accepted: 22 July 2015; Published: 31 July 2015

Abstract: Carbon nanotubes are highly versatile materials; new applications using them are continuously being developed. Special attention is being dedicated to the possible use of multiwall carbon nanotubes in biomaterials contacting with bone. This study describes the response of murine macrophage-like Raw 264.7 cells after two and six days of culture in contact with artificially generated particles from both, ultra-high molecular weight polyethylene polymer and the composite (multiwall carbon nanotubes and ultra-high molecular weight polyethylene). This novel composite has superior wear behavior, having thus the potential to reduce the number of revision knee arthroplasty surgeries required by wear failure of tibial articulating component and diminish particle-induced osteolysis. The results of an *in vitro* study of viability, and interleukin-6 and tumor necrosis factor-alpha production suggest good cytocompatibility, similar to that of conventional ultra-high molecular weight polyethylene.

Keywords: *in vitro* macrophages response; knee replacement; multiwall carbon nanotubes; ultra-high molecular weight polyethylene; wear debris cytotoxicity

---

1. Introduction

Knees carry half of the body weight and provide support and mobility to the human body. In recent years, the incidence of joint degeneration has increased considerably in the young and elderly population; the National Center for Health Science reported that there are more than 300,000 knee replacements per year in the United States [1]. In total knee replacements (TKRs) the production of wear debris is expected because of the sliding and rotating movements of the femoral component against the bearing surface. The articular component material must be completely biocompatible to avoid strong immune reactions due to the interaction of this wear debris with the human body fluids and tissues surrounding the knee joint. Wear debris and degradation products created while using the implant should not cause any inflammatory responses or secondary effects.

Ultra-high molecular weight polyethylene (UHMWPE) has been proven to be a good counterpart material when articulating against cobalt-chromium-molybdenum (Co-Cr-Mo) femoral components in TKRs [2,3]. It displays a very low friction coefficient and it is widely used in the orthopaedic field as a bearing surface in different artificial joints. Because of its high wear resistance and high impact strength, this material remains the material of choice for the fabrication of articular tibial components. Nevertheless, its performance is affected by high creep when compared to metal and bone [4,5]. Also, there has been ongoing concern about the use of UHMWPE because wear of this material often leads to the generation of numerous micron and submicron sized particles (wear debris), which can cause a number of immunological responses and ultimately lead to osteolysis and loosening of the knee implant [4–6].

One of the most researched areas in the last years is the addition of multiwall carbon nanotubes (MWCNT for different applications in the biomedical field; nonetheless, before such material can be incorporated into biomedical devices, the toxicity and biocompatibility of CNTs needs to be thoroughly investigated [7]. The effect of carbon nanotubes on living organisms has been recently studied [7,8]. However, the results of these studies remain controversial [2,3]. Even though carbon nanotubes seem to be a viable option to improve the tribological properties of polymer/composite materials used in orthopedic implants, the biological response of UHMWPE reinforced with carbon nanotubes remains unknown [2,3,9]. This study will focus on the analysis of UHMWPE reinforced with multiwall carbon nanotubes (UHMWPE/CNT) and its cytotoxicity when compared to commercial UHMWPE. The usefulness of UHMWPE/CNT composite must be accompanied by a positive cell interaction to be considered as a viable biomaterial for orthopedic implant consideration.

2. Materials and Methods

The material of interest in this study is the wear debris obtained mechanically from produced samples made from a composite formed by a polymer matrix (UHMWPE) and MWCNT, which is simultaneously compared to the wear debris of samples produced using UHMWPE (orthopedic grade). The multiwall carbon nanotubes are commercial products that were purchased from Sky Spring Nanomaterials, USA. According to the specifications of the manufacturer, the nanotubes are 20–30 nm in diameter and about 10–30 μm in length. Their characteristics, as specified by the manufacturer, are display in Table 1. In literature, one aspect that is considered for MWCNTs to be used in biomedical applications is the purity of this material; this is essential to avoid detrimental reactions when the material is in contact with the human body [8]. Moreover, the XEDS spectrum of the MWCNTs was obtained in a Hitachi S4800 field emission scanning electron microscope (FESEM) and is displayed on Figure 1.

Table 1. Multiwall nanotubes (MWCNTs) general characteristics as specified by the manufacturer.

Multiwall Nanotubes (MWCNTs) Characteristics	
Purity	>95 wt%
Outside diameter	20–30 nm
Inside diameter	5–10 nm
Length	10–30 μm
SSA	>110 m^2/g
Ash	<1.5 wt%
Amorphous carbon	<3.0%
Electrical conductivity	>100 s/cm
Bulk density	0.28 g/cm^3
True density	~2.1 g/cm^3
Manufacturing method	Catalytic CVD

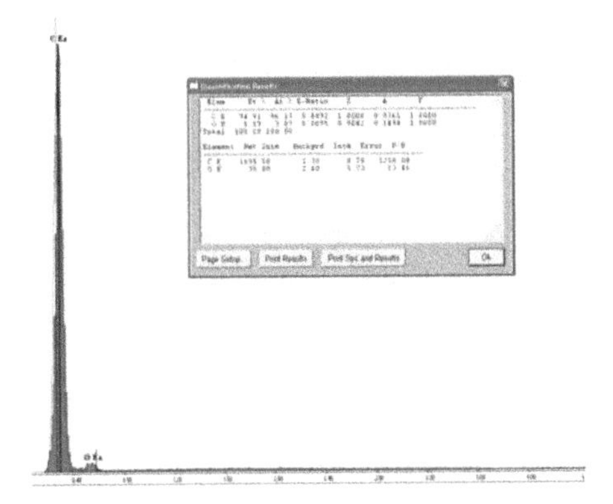

Figure 1. XEDS analysis displaying chemical composition of carbon nanotubes in as-received conditions.

The XEDS analysis did not revealed the presence of any impurities other than oxygen in the carbon nanotubes. Additionally, UHMWPE powder, engineered CNTs and the composite, in a powder form, were tested to observe how the cells reacted to each individual material and the effect of the material size. The materials were tested with the murine macrophage cell line Raw264.7. Macrophages (MΦ) are part of the innate immune system, and their main function is to engulf and digest any cellular debris or pathogen present in the body. The focus of the study was to examine the cell response in the presence of UHMWPE/MWCNTs and determine the cytokine production, specifically, the production of Interleukin-6 (IL-6) and tumor necrosis factor-alpha (TNF-α), pro-inflammatory markers.

2.1. Nanocomposite Material

Medical grade UHMWPE was supplied by Ticona Engineering Polymers, Inc. (Bishop, TX, USA); the trade name of the material used in this study was GUR®1020. The manufacturer specifications of UHMWPE powder were: density = 0.930 g/m^3, mean particle size 137 µm, and average molecular weight = 3.5 million g/mol. The MWCNTs were purchased from Sky Spring Nanomaterials (Houston, TX, USA). The manufacturer specifications of CNTs were: diameter range = 20–30 nm, length of the tubes = 10–30 µm, and purity >95%.

The CNTs were mixed with UHMWPE using the experimental protocol developed Xue *et al.* [10] to produce a homogeneous mixture to be used as the raw material in a compression molding machine. Figure 2 presents a SEM image of the raw composite material, where it can be noticed that the nanotubes are entangled on the UHMWPE particles. The compressed samples of both UHMWPE polymer and UHMWPE/CNT nanocomposites (1.25 wt% of MWCNT) were prepared under optimized testing conditions of pressure, temperature, and time, as determined from our previous studies [10,11].

Figure 2. Micrographs of composite powder at 40 kx.

Hardness is defined as the resistance of a material to penetration; in the case of polymers, hardness is usually measured with a durometer that measures the resistance of the plastic to penetration of a spring loaded needle-like indenter by Shore D scale, in the case hard "plastics". In order to assess the effect of carbon nanotubes on the hardness of UHMWPE, twelve different concentrations were tested, ranging from 0.25 to 20 wt%. From the hardness results, an optimal concentration was chosen for the cytotoxicity studies.

The wear particles from the polymer and the nanocomposite were generated mechanically in the presence of ethanol. After draining the ethanol from the system, the particles were collected and oven-dried, and then sterilized by γ rays at a dose of 25 kGy in a X-Rad 160 irradiatiot. The sterilized particles were used as the raw material in this study. Micrographs of the generated wear particles show that the particles of both materials are in the form of fibrils and have an irregular shape. The size of the particles (0.5–2 μm) was comparable to the phagocyte range for biological response studies associated with secretion of cytokines inductive of osteolysis, as described by Green *et al.* [12], Revell [5], Kurtz [2] and Mcgee [6].

2.2. Dose Response

The control group comprised Raw264.7 macrophage-like cells that were seeded in the wells of 12-well plates at a density of 100,000 cells/well and cultured in the presence of growth medium. Cells were grown in HyClone DMEM hyglucose culture media, 10% fetal bovine serum (FBS), 1% 1-glutamin, 1% of antibiotics (penicillin and streptomycin) and 1% of sodium bicarbonate per container of 500 mL of media. The sterilized particles used in this study (UHMWPE and UHMWPE/CNT particles with an average particle size of 130 μm, engineered MWNTs and, UHMWPE and UHMWPE/CNT debris) were suspended in dimethylsulfoxyde (DMSO) at a concentration of 5 mg/mL, and vortex to ensure a homogenous suspension of the particles immediately prior to pipetting. The concentrations studied ranged from 1 μg/mL to 50 μg/mL for all treatments. Media and campthotecin were used as positive and negative controls and were included in each experiment; also, DMSO was assessed in the dose response analysis, since DMSO was used as the vehicle control. The particles were added to the cells after the 24-h growth period (day 1); then, the culture plate was placed in the incubator for another 24 h. The medium and treatments were refreshed on day 2, and the cell viability was assessed on day 3. All experiments were carried out in triplicates.

Viability was assessed using a luminescent cell viability assay (Cell Titer-Glo Luminescent). The reagent lyses the cells and interacts with the adenosine triphosphate (ATP) within the cells. When the reagent interacts with the ATP, the cells luminesce and produce a signal that is later on detected by the luminometer, where the viability is assessed; the more luminescence, the higher the viability. Briefly, the viability assessment takes place on day 3, the cells are scrapped from the well walls and 100 μL of the cell suspension from each well are transferred to an opaque 96-well plate. Then, 100 μL of cell titer Glo Luminescent are added to each well for the viability assay. Medium was added to three wells in the opaque 96-well plate and were used as blanks. The cells well then placed in the luminometer (Luminoskan Ascent from Labsystems), and the viability was assessed. The voltage used was 900 V and each well was analyzed for 10 s. Once the viability was assessed, the half maximal effective concentration (EC_{50}) was established and used in the acute and chronic exposure studies.

2.3. Acute and Chronic Exposure

In the acute and chronic exposure, viability and pro-inflammatory cytokine production were assessed to establish the macrophage response to the nano-particles (wear debris and carbon nanotubes) and micro-particles (raw material). Lipopolysaccharide (LPS) endotoxin served as a positive control for the activation of the cells, LPS provides the most robust stimulus to macrophages and it attest to the cell viability and reaction in the culture. All assays were run in duplicates.

Viability for the acute exposure was assessed using the same luminescent cell viability assay as in the dose response and the same protocol was followed: the cells were seeded in 12-well plates at a density of 300,000 cells per well on day 0, the particles were added on day 1 and refreshed on day 2 at a concentration of 12.5 μg/mL, the LPS and negative control were also activated on day 3, and the viability was assessed in the luminometer on day 3. On the other hand, for the chronic exposure, cells were seeded on day 0 in 12-well plates at a density of 40,000 cells per well. The particles were added on day 1 and refreshed on days 4 and 6. The LPS and camptothecin controls were activated on day 6 as well, and finally, viability was assessed in the luminometer on day 7.

Pro-inflammatory cytokine production, specifically, TNF-α and IL-6, were assessed using commercially available enzyme-linked immunosorbent assays (ELISA) kits (BioLegend and BD Biosciences, San Diego, CA, USA). The ELISA analyses were performed according to manufacturer's instructions. For the TNF-α production, microtiter ELISA plates were coated with anti-TNFα capture antibody over night at 4 °C. The ELISA kit used to measure the IL-6 production included coated plates with anti-IL-6 capture antibody. The plates were then blocked at room temperature with 3% bovine serum albumin in deionized water. Cell culture supernatants from treated MΦ were added to the plates. Following binding of cytokines to the capture antibodies, the plates were incubated with biotin-conjugated anti-TNFα and IL-6 antibodies, respectively, followed by streptavidin-HRP (Biosource/Life Technologies, Grand Island, NY, USA). The chromogenic substrate 3,3′5,5′-tetramethylbenzidine (TMB) (BD OptEIA, San Jose, CA, USA) was utilized for color development and absorbance was measured by microplate spectrophotometer (VersaMax Microplate Reader, Molecular Devices, Silicon Valley, CA, USA). Cytokine concentrations were calculated against murine recombinant cytokines (BD Pharmingen, San Jose, CA, USA).

3. Results and Discussion

3.1. Hardness Testing

Hardness of pure and reinforced samples was measured with a durometer. In order to understand the effect of carbon nanotubes in the polymer matrix, samples with twelve different concentrations were also fabricated (See Table 2). The load applied on each sample, according to the specifications of the Shore D scale for UHMWPE (ASTM standard D2240), was 4536 g. Figure 3 displays the average hardness value dependig on CNTs concentration. From the graph in Figure 3, it can be concluded that low and high concentrations of CNTs are detrimental for UHMWPE, since these concentrations present the lowest hardness values.

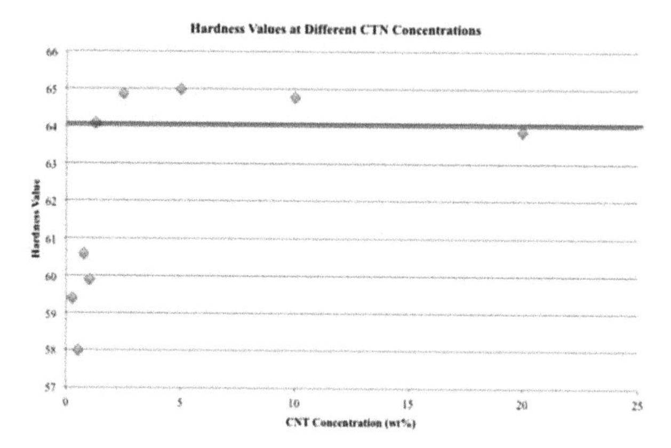

Figure 3. Graph displaying CNT effect on hardness values. The red line stands for the hardness value of the CNT concentration used in this study.

Table 2. Experimental hardness of nanocomposite at different CNT concentrations.

Machine	**Asker Durometer Model EX-D Serial No. 004011 Type D**								
Scale	**Shore D**								
Analysis	**CNT Concentration Effect on Hardness**								
Test date	**11-Jun-12**								
CNT	**Indent**								
Percentage	**1**	**2**	**3**	**4**	**5**	**6**	**7**	**8**	**Average**
0	60	60	62	61	62	62	62	61	61.5
0.25	59	61	60	58	58	59	60	60	59.4
0.50	57	57	58	58	58	59	58	59	58.0
0.75	60	59	61	61	61	61	61	61	60.6
1.00	59	60	60	61	60	60	60	59	59.9
1.25	64	64	64	64	64	65	64	64	64.1
2.50	65	64	65	65	66	64	65	65	64.9
5.00	65	64	64	66	66	65	64	66	65.0
10.00	65	64	64	64	66	64	66	65	64.8
20.00	63	64	64	64	64	65	64	63	63.9

The wear particles from the polymer and the nanocomposite generated mechanically were observed under the scanning electron microscope at 15 kV (SEM-Hitachi TM-1000) and they are displayed in Figure 4. The polymer and nanocomposite particles displayed irregular shapes and variable size. However, the nanocomposite debris was characterized by its fibrillar structure.

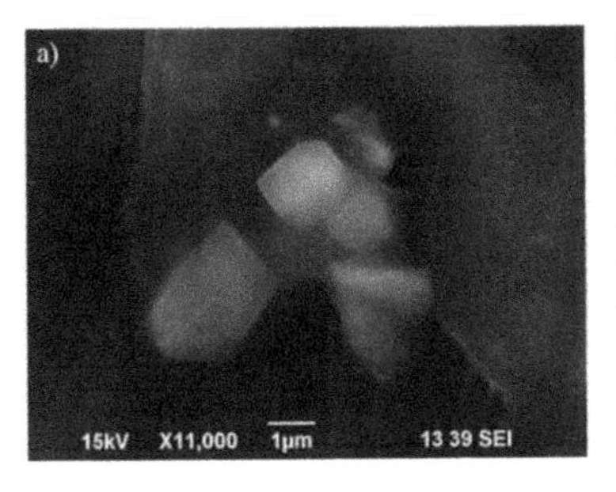

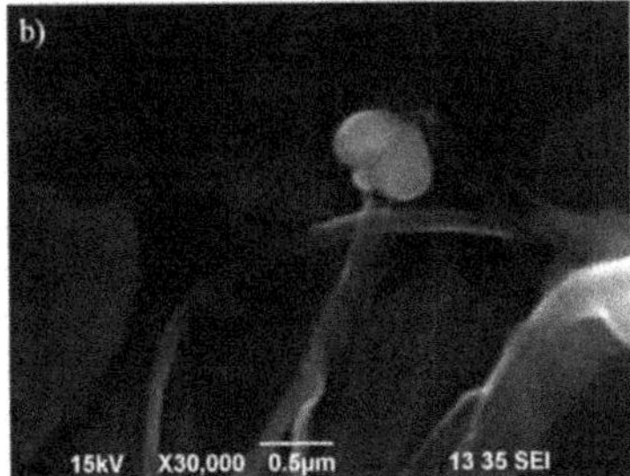

Figure 4. Polymer (**a**) and nanocomposite (**b**) wear debris micrographs.

3.2. Dose Response

The results from the dose response and viability assays showed that UHMWPE micro-particles displayed the lowest EC_{50} (between 6–7 μg/mL) while the UHMWPE wear debris displayed an EC_{50} between 10–15 μg/mL, both of these are way below the EC_{50} from the DMSO, which is the vehicle control. In other words, from the dose response it can be concluded that UHMWPE is cytotoxic in a micro and nano-scale. On the other hand, UHMWPE/CNT in any of its forms did not reach the EC_{50}; because of this, additional concentrations had to be explored. For the composite treatments, 5 additional concentrations were analyzed from 50 to 100 μg/mL. Nevertheless, the EC_{50} for UHMWPE/CNT debris was about 66 μg/mL while the EC_{50} for the UHMWPE/CNT micro-particles could not be reached.

3.3. Acute and Chronic Exposure

In the acute and chronic exposure it was necessary to use a single dose for all the treatments. Moreover, because this is a comparative study between the cytotoxicity of UHMWPE and the UHMWPE/CNT, it was important to assess the cells under the same conditions. It was decided that the dose at which the EC_{50} for the UHMWPE debris was reached, was the most adequate for the assays. Figures 5 and 6 display the results of the viability assays performed for the acute and chronic exposure, respectively.

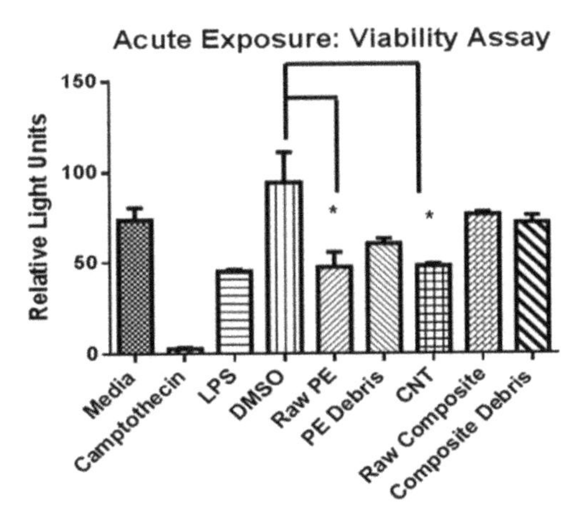

Figure 5. Viability assay for the acute exposure.

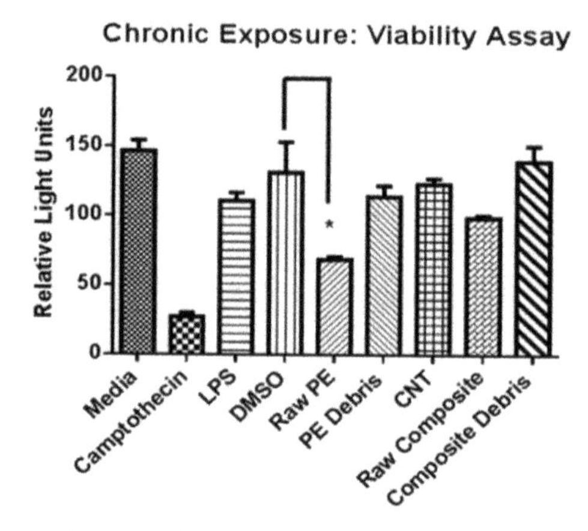

Figure 6. Viability assay for the chronic exposure.

The viability and proliferation assays (acute and chronic exposure) showed no significant differences between the control (DMSO) and UHMWPE/CNT in any of its forms. As a matter of fact, in the chronic exposure, UHMWPE/CNT displayed a viability of 98.5% ± 4.6% (raw material) and 75.3% ± 1.1% (wear debris) while UHMWPE displayed a viability of 86.6% ± 3.4% and 52.2% ± 1.2% for the raw material and the debris, respectively. Comparing viability results in the chronic and acute exposure, it can be concluded that MΦ recover from the initial exposure to UHMWPE in its raw form, since viability when from 50.1% ± 4.5% to 86.6% ± 3.4%. On the other hand, the opposite effect was seen in the UHMWPE wear debris, with this material, the longer the exposure, the lower viability values are seen. The same effect is observed in the presence of UHMWPE/CNT. Nevertheless, the decrement in vitality is less obvious.

Tumor necrosis factor alpha (TNF-α) production for acute and chronic exposure where measured by ELISA and the results are displayed in Figures 7 and 8. TNF-α production in the presence of the particles was compared to the negative control (media). Furthermore, the results of UHMWPE where compared to those obtained for the polymer/composite. All tests were run three times and in duplicates. From the graph in Figure 7 it can be observed that there is a significant difference (39.8%) in the TNF-α production when the cells are exposed to UHMWPE alone and when they are exposed to the reinforced polymer at 48 h. However, when the exposure time increases, the difference between the TNF-α production becomes non-significant. Another conclusion that can be drawn from this analysis is that bigger particles will increase the inflammatory response of the cells.

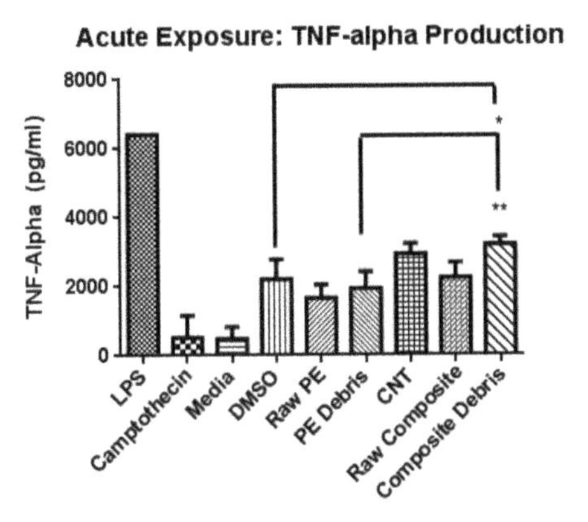

Figure 7. TNF-α production after 48 h of exposure.

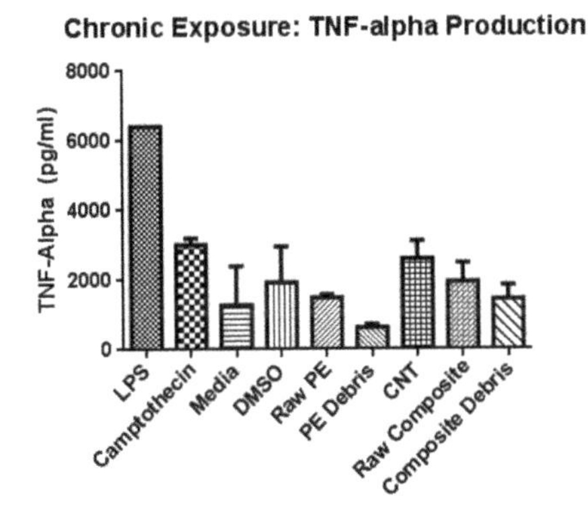

Figure 8. TNF-α production after 168 h of exposure.

Finally, IL-6 production was also assessed to determine the inflammatory response to the cells in the presence of the different particles. Figures 9 and 10 display the graphs with the results obtained in the acute and chronic exposure. Even though UHMWPE stimulated the cells enough for them to produce IL-6, the amount that the MΦ produced is extremely low. Nevertheless, it has to be mentioned that UHMWPE/CNT did not cause any effect on the cells when it referred to the IL-6 production. These results are in agreement with those reported by Chlopek *et al.* [13] where IL-6 production was measured in the presence of polysulfone (PSU) with high purity carbon nanotubes. Additionally, this study measured the potential of macrophages to release free radicals when exposed to carbon nanotubes or carbon particles of similar dimensions [13]. Chlopek *et al.* [13] confirmed a good biocompatibility of carbon nanotubes, characterized by a lack of IL-6 production as well as free radicals induction.

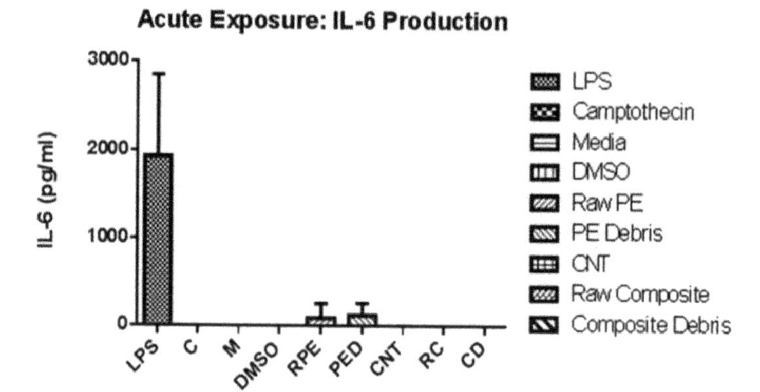

Figure 9. IL-6 production after 48 h of exposure.

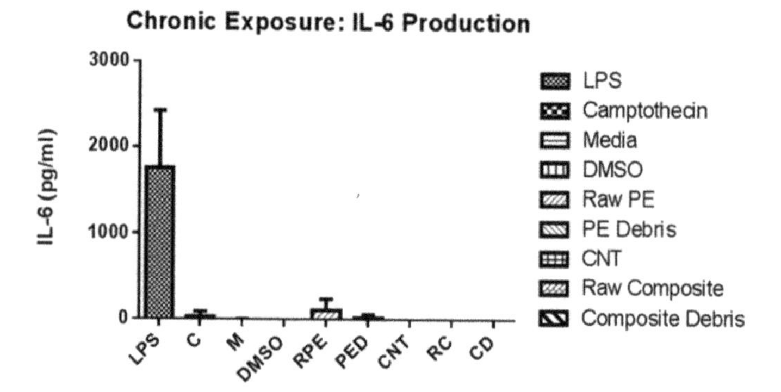

Figure 10. IL-6 production after 168 h of exposure.

Figure 11 displays cells that have been exposed to different particles. Photographs A and B display MΦ exposed to UHMWPE, in its raw and debris form, respectively, while photographs C and D display MΦ exposed to UHMWPE/CNT, also in its raw and debris form, respectively. As it can be observed from the photograph B, the cells seemed to have morphed their appearance more and their pseudopods are larger when compared to those on the other images. Moreover, the cells appeared to be less confluent in presence of UHMWPE debris. For cells exposed to the polymer/composite, the micro size composite seemed to have enlarged the cells when compared to the debris. Nonetheless, the shape of the cells and confluence in the wells appears to be similar.

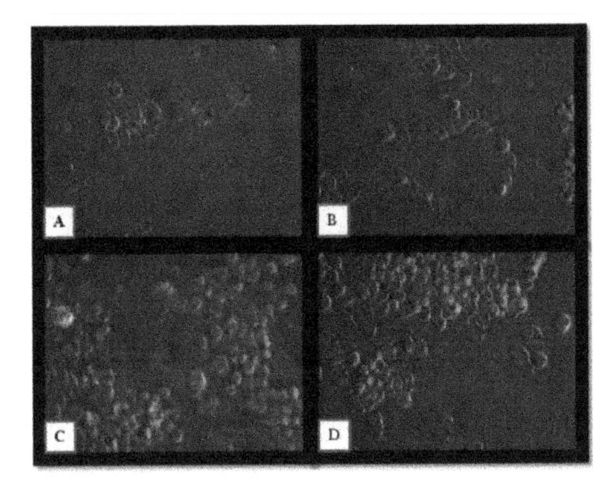

Figure 11. Micrographs of cells after 48 h of exposure to different micro and nano particles (160X). (**A**) Raw UHMPWE, (**B**) UHMWPE debris, (**C**) Raw UHMWPE/CNT, and (**D**) UHMWPE/CNT debris.

3.4. Discussion

Different concentrations of carbon nanotubes in the polymeric matrix were evaluated for this study. Hardness testing showed higher values for UHMWPE/CNT at concentrations above 1 wt% and below 42 wt% CNT than that obtained for pure UHMWPE. The increase in hardness was not very significant from the chosen CNT concentration (1.25 wt%) (2.6 points in the D-Shore scale). However, a small increment in hardness value translates into better creep and wear resistance. It was concluded that the higher the concentration of carbon nanotubes, the higher the values for hardness testing. However, it was noted that after an addition of about 10 wt% or more, the hardness values plateaued. Also it was noted that when the addition of carbon nanotubes exceeded 20 wt%, lower hardness values were obtained.

Viability, and tumor necrosis factor alpha (TNF-α) and interleukin-6 (IL-6) production were assessed by macrophage Raw264.7 cells. The results of this *in vitro* study suggest good cytocompatibility of the UHMWPE/CNT material when compared to UHMWPE (current commercial material); these results are in agreement with those reported by Reis *et al.* [8] in his studies with osteoblasts. Viability assays demonstrated that in an initial exposure (acute study), macrophages react strongly to particles with the micron and sub-micron size particles (debris). However, when the cells stay in the presence of the material, bigger size particles (raw material) decreases the viability. Even though there were no significant differences between the cells viability when exposed to UHMWPE than when exposed to UHMWPE/CNT, cell viability was slightly higher in the presence of UHMWPE/CNT, especially in its debris form. On the other hand, when pro-inflammatory cytokine production, specifically TNF-α and IL-6, was measured, it was noted that a while IL-6 was barely present in the supernatants obtained after 48 and 168 h of exposure to the particles, TNF-α levels were relatively high, especially for the polymer/composite debris after only 48 h of exposure. Nonetheless, TNF-α levels stabilized themselves and no significant difference can be observed among UHMWPE, polymer/composite, and controls.

The use of carbon nanotubes in the medical field has been questioned in the last years because of the controversial results of the biocompatibility of CNTs. However, in this study, the lack of presence of IL-6 in the composite group suggests good cytocompatibility, since IL-6 is a powerful inductor of bone resorption through osteoclast activation. The use of this novel composite in the orthopaedic field could benefit patients in different ways. The composite has demonstrated to be not only superior in mechanical properties, but also it seems that the use of this material could decrease the number of total knee replacement failures due to osteolysis.

4. Conclusions

The cellular response associated with engineered wear debris from UHMWPE (current commercial material for tibial articular inserts in TKRs) was compared to the cellular response associated with wear debris from UHMWPE/CNT samples. Murine macrophage-like cells (Raw264.7) were used to assess the cytotoxicity of these materials. A dose response was first performed to establish the EC_{50} to be used in the acute and chronic exposure to the micro and nanoparticles. Viability was assessed through a luminescent cell viability assay and the cytokines production, specifically TNF-α and IL-6, were assessed by ELISA.

The viability and pro-inflammatory cytokine production was assessed by macrophage-like cells Raw264.7. The main purpose of this study was to compare the cytotoxicity of UHMWPE and UHMWPE/CNT. In the first stage of this research, a dose response assessment was performed to calculate the half maximal effective concentration (EC_{50}) to be used in the acute and chronic exposure. The dose response was assessed through the viability of the cells with a commercial luminescent cell viability assay. The results of the dose response display a higher EC_{50} for UHMWPE/CNT in both forms (raw material and artificially generated debris) than that found for UHMWPE.

The viability assessment in an acute and chronic exposure displayed similar results. There were no significant difference between UHMWPE and the composite or the controls and the composite. However, UHMWPE in its raw form displayed a significant difference when compared to the vehicle control (DMSO). This suggests that the presence of carbon nanotubes in the polymer matrix actually protects the cells from the plastic and increases the viability of MΦ.

On the other hand, pro-inflammatory cytokine production assessment showed different results. UHMWPE/CNT displays a higher production of TNF-α than UHMWPE, especially in micron and sub-micron size (wear debris) in an initial exposure (acute). Nevertheless, the chronic exposure revealed similar levels of TNF-α production in controls, UHMWPE, and polymer/composite. It was noted that in the chronic exposure, bigger particles (raw materials) elicit a higher cytokine production. Finally, the production of IL-6 was assessed by ELISA. From the different materials studied, UHMWPE was the only one that produced IL-6. However, the IL-6 production was so low, that it was not significant when compared to the controls.

Acknowledgments: The authors recognize the generous funding contributions of National Council for Science and Technology (CONACyT) to this research effort. This work was also supported by the National Institutes of Health—Research Center at Minority Institutions (5G12RR008124). Kristina Barron was funded by the NIH—Marcus U-Star program (T34-GM008048) and Raquel Suro was funded by the NIH—Rise program for graduate students (R25 GM 069621). The authors would like to thank the UTEP-RCMI Border Biomedical Research Center Core Facilities: Cytometry, Screening, and Imaging; and Biomolecule Analysis.

Author Contributions: Nayeli Camacho wrote the paper, designed the material and performed its mechanical and biological characterization under the supervision of Stephen W. Stafford and Kristine M. Garza. Additionally, Kristine M. Garza coordinated and supervised the cytotoxicity experiments and contributed to the data analysis. Raquel Suro and Kristina I. Barron contributed to the cytotoxicity experiments and data analysis.

Conflicts of Interest: The authors declare no conflict of interest.

References

1. Janeway, P.A. Bioceramics: Materials That Mimic Mother Nature. *Am. Ceram. Soc. Bull.* **2006**, *8*, 26–30.
2. Kurtz, S.M. *The UHMWPE Handbook: Ultra High Molecular Weight Polyethylene in Total Joint Replacement*; Elsevier Inc.: San Diego, CA, USA, 2004.
3. Kurtz, S.M. *UHMWPE Biomaterials Handbook*, 2nd ed.; Elsevier Inc.: Burlington, MA, USA, 2009.
4. Schmalzried, T.P.; Callaghan, J.J. Current Concepts Review: Wear in Total Knee and Hip Replacements. *J. Bone Jt. Surg.* **1999**, *81*, 115–132.
5. Revell, P.A. The Combined Role of Wear Particles, Macrophages and Lymphocytes in the Loosening of Total Joint Replacements. *J. R. Soc. Interface* **2008**, *5*, 1263–1278. [CrossRef] [PubMed]

6. Mcgee, M.A.; Howie, D.W.; Costi, K.; Haynes, D.R.; Wildenauer, C.I.; Pearcy, M.J.; McLean, J.D. Implant Retrieval Studies of the Wear and Loosening of Prosthetic Joints: Review. *Elsevier Sci. Wear* **2000**, *241*, 158–165. [CrossRef]
7. Smart, S.K.; Cassady, A.I.; Lu, G.Q.; Martin, D.J. The Biocompatibility of Carbon Nanotubes. *Carbon* **2006**, *44*, 1034–1047. [CrossRef]
8. Reis, J.; Kanagaraj, S.; Fonseca, A.; Mathew, M.T.; Capela-Silva, F.; Potes, J.; Pereira, A.; Oliveira, M.S.; Simões, J.A. *In vitro* Studies of Multiwall Carbon Nanotube/Ultrahigh Molecular Weight Polyethylene Nanocomposites with osteoblast-like MG63 Cells. *Braz. J. Med. Biol. Res.* **2010**, *43*, 476–482. [CrossRef] [PubMed]
9. Gomez-Barrena, E.; Puertolas, J.A.; Munuera, L.; Konttinen, Y.T. Update on UHMWPE Research: From the Bench to the Bedside. *Acta Orthop.* **2008**, *79*, 832–840. [CrossRef] [PubMed]
10. Camacho, N. Characterization of Ultra-High Molecular Weight Polyethylene (UHMWPE) Reinforced with Multiwall Carbon Nanotubes for Orthopedic Applications. Ph.D. Dissertation, Materials Science Engineering, University of Texas at El Paso, El Paso, TX, USA, 4 January 2013.
11. Green, T.R.; Fisher, J.; Stone, M.; Wroblewskid, B.M.; Ingham, E. Polyethylene particles of a "critical size" are necessary for the induction of cytokines by macrophages *in vitro*. *Biomaterials* **1998**, *19*, 2297–2302. [CrossRef]
12. Yang, X.; Wu, W.; Jacobs, O.; Schädel, B. Tribological behavior of UHMWPE/HDPE blends reinforced with Multi-wall Carbon nanotubes. *Polym. Test.* **2006**, *25*, 221–229.
13. Chlopk, J.; Czajkowska, B.; Szaraniec, B.; Frackowiak, E.; Szostak, K.; Béguin, F. *In Vitro* Studies of Carbon Nanotubes Biocompatibility. *Carbon* **2006**, *44*, 1106–1111.

MDPI
St. Alban-Anlage 66
4052 Basel
Switzerland
Tel. +41 61 683 77 34
Fax +41 61 302 89 18
www.mdpi.com

Lubricants Editorial Office
E-mail: lubricants@mdpi.com
www.mdpi.com/journal/lubricants

CPSIA information can be obtained
at www.ICGtesting.com
Printed in the USA
LVHW021723220323
742314LV00038B/1095